RESISTANCE, RESILIENCE, AND RECOVERY FROM DISASTERS

COMMUNITY, ENVIRONMENT AND DISASTER RISK MANAGEMENT

Series editor: William L. Waugh, Jr.

Urban RISK Reduction: An Asian Perspective, Volume 1

Water Communities, Volume 2

Integrated Lagoon Fisheries Management: Resource Dynamics and Adaptation, Volume 3

Climate Change Adaptation and Disaster Risk Reduction: Issues and Challenges, Volume 4

Climate Change Adaptation and Disaster Risk Reduction: An Asian Perspective, Volume 5

Climate and Disaster Resilience in Cities, Volume 6

Disaster Education, Volume 7

Droughts in Asian Monsoon Region, Volume 8

Environment Disaster Linkages, Volume 9

Community-based Disaster Risk Reduction, Volume 10

Climate Change Modeling for Local Adaptation in the Hindu Kush-Himalayan Region, Volume 11

Ecosystem-based Adaptation, Volume 12

Water Insecurity: A Social Dilemma, Volume 13

Risks and Conflicts: Local Responses to Natural Disasters, Volume 14

Building Resilient Urban Communities, Volume 15

Hyogo Framework for Action and Urban Disaster Resilience, Volume 16

Local Disaster Risk Management in a Changing Climate: Perspective from Central America, Volume 17

Recovering from Catastrophic Disaster in Asia, Volume 18

The Tourism–Disaster–Conflict Nexus, Volume 19

Improving Flood Management, Prediction and Monitoring, Volume 20

COMMUNITY, ENVIRONMENT AND DISASTER RISK MANAGEMENT VOLUME 21

RESISTANCE, RESILIENCE, AND RECOVERY FROM DISASTERS: PERSPECTIVES FROM SOUTHEAST ASIA

EDITED BY

MA. REGINA M. HECHANOVA, PH.D.
Ateneo de Manila University

LYNN C. WAELDE, PH. D.
Palo Alto University

emerald PUBLISHING

United Kingdom – North America – Japan
India – Malaysia – China

Emerald Publishing Limited
Howard House, Wagon Lane, Bingley BD16 1WA, UK

First edition 2020

British Library Cataloguing in Publication Data
A catalogue record for this book is available from the British Library

ISBN: 978-1-83909-791-1 (Print)
ISBN: 978-1-83909-790-4 (Online)
ISBN: 978-1-83909-792-8 (Epub)

ISSN: 2040-7262 (Series)

Printed and bound by CPI Group (UK) Ltd, Croydon, CR0 4YY

ISOQAR
REGISTERED
Certificate Number 1985
ISO 14001

ISOQAR certified Management System, awarded to Emerald for adherence to Environmental standard ISO 14001:2004.

INVESTOR IN PEOPLE

ENDORSEMENTS

Crises and catastrophes are on the rise, becoming more the rule than the exception. Thus, a volume devoted to issues of resistance, resilience, and recovery could not be timelier or more important. Hechanova's and Waelde's insightful work should be required reading for leaders, scholars, and community members globally looking for culturally sound and effective evidence-based intervention frameworks.

Kathleen M. Sutcliffe
Bloomberg Distinguished Professor, Johns Hopkins University and co-author of *Still Not Safe: Patient Safety and the Middle Managing of American Medicine*

Thoughtful and sensitive. The book does not only offer keen ideas about mental health and psychosocial support in times of disaster, it also gives us a deeper appreciation of the concepts of resistance, resilience, and recovery from disasters that is specific to the context that is Southeast Asia. In the midst of the global pandemic we are currently facing, this book gives us valuable insights as to how we can ensue collective resilience among our people in Southeast Asia and move forward in a way that is fit to the context of our new normal. What I appreciate most about the book is that it's giving a space for one of those groups that is often neglected in times of disaster — the disaster responders.

Marshaley J. Baquiano,
PhD University of the Philippines Visayas

Disasters have profoundly affected the evolution of human cultures through known history, and no region has felt these effects more than Southeast Asia (SEA). This volume provides a concise and unique resource for understanding how stress-resistance in the face of disaster, resilience to its most acute effects, and longer-term adaptive recovery are entwined with the cultural ecologies of those affected. These are perhaps the most fundamentally important distinctions for understanding how we can accurately anticipate and respond effectively to the mental health and psychosocial support needs of communities and organizations across SEA impacted by such catastrophic events. I recommend this book as a practical and scholarly resource for informing culturally sensitive studies and humanitarian interventions in general, and especially in SEA context.

Gil Reyes, Ph.D.
Disaster Relief Chair, APA Division of Trauma Psychology

CONTENTS

III. MENTAL HEALTH AND PSYCHOSOCIAL SUPPORT INTERVENTIONS

IV. CONCLUSION

EDITOR BIOGRAPHIES

Ma. Regina M. Hechanova, Ph.D., is a Full Professor and former Chairperson of the Department of Psychology, Ateneo de Manila University (ADMU). Gina was also the past President of the Psychological Association of the Philippines (PAP) and created the Special Interest Group for Mental Health and Psychosocial Support (MHPSS). She was a Lead Researcher in the development of *Katatagan,* a resilience program for disaster survivors and the Editor of the *Philippine Journal of Psychology* Special Edition on MHPSS. She designed an undergraduate course on Disaster and Mental Health and co-created a minor on Humanitarian Aid in ADMU. She was named Outstanding Women in Nation's Service in 2010 and was named 2018 Outstanding Professional in Psychology by the Professional Regulatory Commission. Her research and practice are in the areas of mental health and psychosocial support post-disaster and community mental health interventions in the Philippines.

Lynn C. Waelde, Ph.D., is a Professor in the Psychology Department of Palo Alto University and an Adjunct Clinical Professor in the Department of Psychiatry and Behavioral Sciences of Stanford University School of Medicine. Her research and clinical interests include mindfulness and meditation interventions and provider training, development and adaptation of disaster mental health interventions, diagnosis and assessment of trauma-related issues, and racism-related stress. She is the Founder and Director of the Inner Resources Center, which is a center for mindfulness and meditation intervention, training, and research.

CONTRIBUTOR BIOGRAPHIES

Avegale Acosta is a Lecturer at the Ateneo de Manila University, Psychology Department and a Counseling Psychologist. She specializes in the areas of anxiety, depression, trauma, and grief. She conducts classes, trainings, and workshops that help equip individuals how to respond and recover from critical incidents like disasters.

Daniela Blecha is a development and humanitarian affairs professional. She has worked at several international and non-governmental organizations in Vienna, New York, and Brussels, including the International Organization for Migration, the United Nations Operations, and Crisis Centre, the International Centre for Migration Policy Development and SOS Children's Villages International.

Charlotte Kendra Gotangco is currently the Program Manager for Climate Change and Disaster Risk of the Ateneo Institute of Sustainability and the Chair of the Dept. of Environmental Science of the Ateneo de Manila University. She is working on applying systems thinking approaches to understanding the drivers of risk and mainstreaming resilience planning.

Johnrev Guilaran is an Assistant Professor of psychology at the University of the Philippines Visayas. He is an Integrated Research on Disaster Risk (IRDR) Young Scientist and an International Association of Applied Psychology (IAAP) Task Force on Terrorism and Peacebuilding collaborator. He has a Ph.D. from Massey University, Wellington.

Andrew G. Heise is a PhD candidate in clinical psychology at Palo Alto University and a Lecturer at California State University, Stanislaus. He is interested in trauma, burnout among healthcare providers, psychological first aid, and race-related stress.

Roman Hoffmann is an Environmental Economist at the Potsdam Institute for Climate Impact Research and the Wittgenstein Centre for Demography and Global Human Capital in Vienna. His work focuses on disaster resilience, climate change adaptation and mitigation, environmental migration, and environmental health.

Elirozz Carlie Labaria is a Disaster Risk Reduction and Climate Change Adaptation Specialist of the Ateneo Institute of Sustainability. Her expertise is on DRR and CCA planning and governance mechanisms.

Jason O. Manaois is a Lecturer of Psychology at Xavier University – Ateneo de Cagayan. He took his PhD in social-organizational psychology at the Ateneo de Manila University. He engages in mental health, psychosocial support and community interventions, and organization research.

Hong An Nguyen is a Counselling Practitioner and the Head of the Department of Psychology at Hoa Sen University. He has a master's degree in Psychology from Massey University, Auckland. He is a Member of the EMDR Asia and the Asian Association of Social Psychology.

Adriana Panting is a clinical psychology Ph.D. candidate at Palo Alto University expected to graduate in June 2020. She works as an Anti-Sexual Trafficking Specialist/Program Manager at a trauma counseling agency, Tri-Valley Haven. Her research and clinical interests include mindfulness, ethnoracially marginalized groups, serious mental illness, and trauma.

Grant J. Rich, Ph.D., is a Senior Faculty at Walden University. A Fellow of the American Psychological Association and four of its divisions, he is Lead Editor of six international psychology books, including *Pathfinders in International Psychology, Human Strengths and Resilience, Teaching Psychology Around the World,* and *Internationalizing the Teaching of Psychology*.

Pinky Rose Sabile, MA, is a Lecturer at the Department of Psychology, Ateneo de Manila University and the Director for Human Resource Development at the Ateneo Center for Organization Research and Development (Ateneo CORD).

J. Sedfrey S. Santiago is an Associate Professor at the Ateneo de Manila University where he teaches law courses at the John Gokongwei School of Management, and The Creative Professional in the Fine Arts Department. Among others, he has co-written as lead author articles on disaster published in scholarly journals.

Skultip (Jill) Sirikantraporn, Psy.D., is a Licensed Psychologist in California and New York and is on founding faculty at Fulbright University Vietnam, developing the psychology major and teaching various interdisciplinary undergraduate courses. She is a co-editor and contributor of an international psychology book, *Human Strengths and Resilience: Developmental, Cross-Cultural, and International Perspectives.*

Chantal Ellis S. Tabo-Corpuz is a licensed Psychologist. She is practicing as a school psychologist in Ateneo de Manila University and is a PhD student in Clinical Psychology in the same university. She is interested in mental health prevention and intervention programs, and advocates for its development in the Philippines.

Mendiola Teng-Calleja, Ph.D., is an Associate Professor at the Department of Psychology, Ateneo de Manila University and a Director at the Ateneo Center for Organization Research and Development (Ateneo CORD).

Alicia N. Torres is a clinical psychology Ph.D. student at Palo Alto University. Her research interests include mindfulness and meditation interventions, serious mental illness, and psychosocial rehabilitation.

Angelique Pearl Virtue Villasanta, has a Masters' Degree in Counseling Psychology and is a Lecturer at the Department of Psychology, Ateneo de Manila University.

LIST OF TABLES

LIST OF FIGURES

PREFACE

As of this writing, more than a quarter of a million people are in evacuation shelters as a result of Taal Volcano in the Philippines, which explosively erupted just days ago (National Disaster Risk Reduction and Management Council, 2020). Since then, the region has experienced more than 900 volcanic earthquakes, raising the prospect of volcanic tsunamis. The Philippine Institute of Volcanology and Seismology (PHIVOLCS) has posted an Alert Level 4 for Taal, meaning that a hazardous eruption is possible within hours to days (PHIVOLCS, 2020). The area is covered in feet of volcanic ash, obliterating homes, livelihoods, and communities.

It is now clear that human-induced climate change affects the likelihood and severity of many types of natural disasters (National Academies of Sciences, Engineering, and Medicine, 2016). The reality of climate change and its relationship to increased risk of natural disasters means that disaster preparedness and response must be a priority, as this elevated risk is associated with increasing vulnerabilities worldwide (Van Aalst, 2006). Much work already documents that climate change and disaster have deleterious effects on mental health, particularly for those who are impoverished and marginalized (Hayes, Blashki, Wiseman, Burke, & Reifels, 2018). Thus, disaster preparedness and response must address mental health and psychosocial support (MHPSS), that is, interventions that can address the range of severity of disaster stress sequelae from normative responses to more severe and lasting psychopathological outcomes. The chapters in this volume document that Southeast Asia is particularly vulnerable to disasters of all kinds, which exact a terrible toll of suffering and destruction.

Three key concepts guide the considerations of disaster mental health presented in this volume: resistance, resilience, and recovery, as described in the Johns Hopkins model. Resistance is a form of psychological immunity to developing manifestations of distress and dysfunction following disasters. Resilience refers to the capacity to effectively rebound from distress and dysfunction that occurs as a result of disaster. Recovery refers to the capacity to restore adaptive functioning following distress, impairment, and dysfunction that occur as a result of a disaster (Kaminsky, McCabe, Langlieb, & Everly, 2007). Addressing these three capacities of resistance, resilience, and recovery requires a solid evidence base to guide effective planning and practice that has been informed by all levels – from the individual to the population – and is grounded in knowledge about applications that are culturally appropriate.

The chapters in this volume address ways to enhance resistance, resilience, and recovery using leadership models and MHPSS interventions that are developed and implemented using a whole community approach that is grounded in and guided by cultural considerations. Because much work about disaster planning and recovery is based on work done in Western cultures, in the first chapter, we address cultural considerations for MHPSS in Southeast Asia. Part II addresses leadership and organizational models for MPHSS. The powerful role of different forms of education to reduce disaster vulnerabilities is reviewed in Chapter 2 authored by Roman Hoffmann and Daniela Blecha. In Chapter 3, Mendiola Teng-Calleja, Pinky Rose Sabile, and Angelique Pearl Virtue Villasanta review ways to structure workplace organization to enhance employee and organizational resilience in the face of disasters. In Chapter 4, Elirozz Carlie Labaria, Avegale Acosta, and Charlotte Kendra Gotangco explain how the disaster planning process, including prevention, mitigation, and preparedness efforts, can be integrated with mental health efforts.

Part III of this volume reviews different forms of MHPSS support interventions as they have been applied in Southeast Asian countries. In Chapter 5, Jason O. Manaois, Chantal Ellis S. Tabo-Corpuz, and Andrew G. Heise review psychological first aid and its application in SEA. The next two chapters address specific disaster intervention approaches, namely, mindfulness and art therapy. In Chapter 6, Adriana Panting, Andrew G. Heise, Ma. Regina M. Hechanova, and Lynn C. Waelde review the rationale, cultural adaptations, and evidence base for integrating mindfulness and meditation into MHPSS. In Chapter 7, J. Sedfrey S. Santiago discusses art therapy as a way to allow expression of disaster stress experiences and provide a pathway to financial support for disaster survivors, using the Banglos fisherfolk as an illustration. Disaster responders bear heavy stress exposure burdens. In Chapter 8, Johnrev Guilaran and Hong An Nguyen describe MHPSS interventions for disaster responders themselves. The last chapter of Section II, Chapter 9, authored by Grant J. Rich and Skultip (Jill) Sirikantraporn, addresses posttraumatic growth as an outcome of disaster exposure. In the final chapter in Section IV, Chapter 10, we summarize the challenges and prospects for promoting resistance, resilience, and recovery in SEA.

The chapters in this volume address both the vulnerabilities and protective factors in SEA, highlighting the extent to which geography, poverty, and lack of resources, capability, and evidence-based disaster interventions make SEA populations more vulnerable to disasters and their negative outcomes. On the other hand, the strong sense of family, community, and spirituality appear to be protective factors. The chapters also highlight gaps in MHPSS responses and the need for rigor in examining the impact of interventions.

The idea for this volume developed from our collaboration to develop disaster MHPSS programs for the Philippines following Typhoon Haiyan. We are grateful to our collaborators in that initial effort from Ateneo de Manila University, the Psychological Association of the Philippines, and University of the Philippines Visayas Tacloban College. The funding for the original project that brought us all together was provided by Ateneo de Manila University and Palo Alto University. We also wish to acknowledge the Rockefeller Foundation Bellagio Center Residency Program that funded the initial conceptualization of this book project. We thank Alicia N. Torres for providing editorial assistance in the preparation of this book.

In a region that is constantly challenged by disasters, we hope to highlight how important it is to understand cultures and the possible ways in which the world can learn from Southeast Asia. We hope to encourage further research and translation into disaster mental health practice.

Ma. Regina M. Hechanova, Ph.D.
rhechanova@ateneo.edu

Lynn C. Waelde, Ph.D.
lwaelde@paloaltou.edu
January, 2020

REFERENCES

Hayes, K., Blashki, G., Wiseman, J., Burke, S., & Reifels, L. (2018). Climate change and mental health: Risks, impacts and priority actions. *International Journal of Mental Health Systems*, *12*, 28. https://doi.org/10.1186/s13033-018-0210-6

Kaminsky, M., McCabe, O. L., Langlieb, A. M., & Everly, G. (2007). An evidence-informed model of human resistance, resilience, and recovery: The Johns Hopkins' outcome-driven paradigm for disaster mental health services. *Brief Treatment and Crisis Intervention*, *7*(1), 1–11. https://doi.org/10.1093/brief-treatment/mhl015

National Academies of Sciences, Engineering, and Medicine. (2016). *Attribution of Extreme Weather Events in the Context of Climate Change*. Washington, DC: The National Academies Press. Retrieved from https://doi.org/10.17226/21852

National Disaster Risk Reduction and Management Council. (2020). NDRRMC update: Situational report no. 37 re Taal Volcano eruption. Retrieved from http://www.ndrrmc.gov.ph/attachments/article/4007/Update_on_SitRep_No_37_re_Taal_Volcano_Eruption_issued_on_24Jan2020_6AM.pdf

Philippine Institute of Volcanology and Seismology (PHIVOLCS). (2020). Taal Volcano advisory: 22 January 2020 04:00 pm. Retrieved from https://www.phivolcs.dost.gov.ph/index.php/taal-volcano-bulletin-menu/9679-taal-volcano-advisory-22-january-2020-04-00-pm

Van Aalst, M. K. (2006). The impacts of climate change on the risk of natural disasters. *Disasters*, *30*, 5–18. https://doi.org/10.1111/j.1467-9523.2006.00303.x

PART I

INTRODUCTION

CHAPTER 1

CULTURAL IMPLICATIONS FOR THE PROVISION OF DISASTER MENTAL HEALTH AND PSYCHOSOCIAL SUPPORT IN SOUTHEAST ASIA

Ma. Regina M. Hechanova, Lynn C. Waelde and Alicia N. Torres

ABSTRACT

Southeast Asia (SEA) is a region highly susceptible to earthquakes, volcanic eruptions, and tsunamis, though the region has been underrepresented in disaster mental health research. This chapter addresses risk factors for SEA, including its disaster-prone location, the psychological toll of frequent disasters, and stigma and shame and lack of psychoeducation about psychological help-seeking. Collectivism, strong family ties, and religious faith are among SEA's resilience factors. Culture should be heavily accounted for in mental health and psychosocial support (MHPSS), considering the wide array of cultural differences in spirituality, affect and expression, power distance, and gender and masculinity in SEA. Because culture affects treatment satisfaction, treatment engagement, and treatment outcomes, future research should explore how aspects of SEA culture impact accessibility and engagement in MHPSS.

Keywords: Mental health and psychosocial support; Southeast Asia; culturally adapted interventions; trauma; disaster mental health; culture

---

Resistance, Resilience, and Recovery from Disasters: Perspectives from Southeast Asia
Community, Environment and Disaster Risk Management, Volume 21, 3–13

ISSN: 2040-7262/doi:10.1108/S2040-726220200000021001

The Southeast Asia (SEA) region consists of two geographic regions. Mainland SEA consists of Myanmar, Thailand, Malaysia, Laos, Cambodia, and Viet Nam. Maritime SEA consists of Indonesia, Brunei, Singapore, East Timor, Christmas Island and Cocos Island. It is considered as one of the most disaster-prone regions in the world (United Nations International Strategy for Disaster Reduction (UNISDR), 2010) because it lies along the Pacific Ring of Fire.

Despite its vulnerability, the region is underrepresented in terms of disaster mental health research. In this chapter, we summarize the literature about the impact of disasters, risk factors, and vulnerabilities in SEA. We also describe various aspects of SEA culture that may shape the provision of mental health and psychosocial support (MHPSS) in the region. Culture is a set of behavioral norms and cognitions representative of people from a definable population that are different from members of other populations (Lehman, Chiu, & Schaller, 2004). There is a growing need for cultural adaptations of MHPSS interventions, because most approaches were developed in Western, individualistic cultures with uncertain application in SEA. Culture can impact treatment satisfaction, engagement in treatment, and treatment outcomes (Constantine, 2002), making cultural considerations for MHPSS a vital concern.

The SEA region is located between two great oceans – the Pacific Ocean and the Indian Ocean and in the intersection of geologic plate creating earthquakes, volcanic eruptions, and tsunamis. Not surprisingly, countries in SEA have a history of disasters including typhoons, floods, tsunamis, volcanic eruptions, earthquakes, landslides, epidemics, and droughts (UNISDR, 2010). Between 1981 and 2010, the region experienced as many 912 disasters such as floods (47%) and tropical cyclones (38%) (Hechanova & Waelde, 2017). During this period, disasters have killed 199,075 and affected 310,443,666 survivors in the region (Samphantharak, 2014).

PSYCHOLOGICAL IMPACT OF DISASTERS IN SEA

There is an extensive literature demonstrating that disasters' survivors experience depressive symptoms and related symptoms of mood disorders including fatigue, loss, helplessness, withdrawal, and enduring grief reactions. In addition, the experience of trauma exacerbates previous disorders such as substance use disorders and depression (Bonanno, Brewin, Kaniasty, & La Greca, 2010). Studies in SEA have reported similar symptoms among disaster survivors. For example, a post-Typhoon Haiyan study in the Philippines described somatic, emotional, cognitive, and behavioral symptoms similar to that reported by disaster survivors in other parts of the world (Hechanova, Ramos, & Waelde, 2015). Beyond these deleterious outcomes, survivors of the Mount Merapi volcanic eruption in Indonesia (Warsini, Buettner, Mills, West, & Usher, 2014) and the survivors of the 2004 tsunami in Thailand (Rigg, Grundy-Warr, Law, & Tan-Mullins, 2008) described a sense of *solastalgia* or the loss of a feeling of safety and comfort because of the devastation of their environment.

Around the world, studies suggest that about 30% of disaster survivors experience transient PTSD symptoms but only about 5%–10% of survivors develop full-blown PTSD (Bonanno et al., 2010). Although there is a lack of epidemiological studies in SEA from which to draw conclusions, initial reports from Asia report a higher proportion (8.6%–57.3%) of survivors with PTSD symptoms immediately after a disaster. However, in the medium term, by nine months to a year after the disaster, this proportion decreases to 2.3%–32%, and two years after the disaster, it deceases to 1.2%–7.6% (Udomratn, 2008). Although the relatively higher prevalence may be because of small and biased samples (Bonanno et al., 2010), the percentage of those who are at risk for PTSD may also be due to the lack of mental health resources, poor disaster preparedness, social and educational disruption, and enduring poverty (Dawson et al., 2014). For example, a study in the Philippines after Typhoon Haiyan reported that millions of survivors were displaced and did not have adequate shelter for months (Hechanova, Ramos, et al., 2015). A study in Thailand showed that among displaced survivors of the 2004 tsunami, 12% had PTSD symptoms as compared to 7% of non-displaced survivors (van Griensven et al., 2006).

In addition to the classical symptoms of trauma, there appear to be some cultural nuances of trauma in SEA. Tuliao's (2014) review of the Philippine literature concluded that Filipinos do not differentiate between physical and mental disorders. He cited a study by Shakman (1969) that described how survivors seek out indigenous and folk healers because of disturbed behavior and somatic complaints that appear to have no medical causes. The somatization of illness may not be unique to the Philippines, however, because different cultural groups somaticize psychological symptoms (Chhim, 2013). There is also evidence of some trauma-related symptoms that appear to be culturally specific. In Cambodia, the term *baksbat* (broken courage) is used to describe posttrauma symptoms including feeling fearful, mute and deaf, and lacking trust in others (Chhim, 2013).

MENTAL HEALTH PROFESSIONALS

Fortunately, countries in SEA have increasingly instituted disaster risk reduction and management (DRRM) and mental health policies and programs. Indonesia, Myanmar, and Thailand included MHPSS in their disaster preparedness programs (Ito, Setoya, & Suzuki, 2012). A barrier to the delivery of MHPSS is the lack of mental health resources to provide the needed services. Countries in SEA (Lao, Cambodia, Myanmar, Indonesia, Viet Nam, and the Philippines) have the lowest number of psychiatrists in Asia (Ito et al., 2012). Fortunately, a protective factor is the active role of non-governmental organizations (NGOs) and universities in providing psychosocial rehabilitation services in the community. Some countries such as Indonesia have begun to embed community mental health in their nursing training programs (Ito et al., 2012).

In addition, although there may be dearth of formal mental health care providers, a protective factor in the region is the existence of traditional healers.

Many Cambodians seek help from herbalists (*Kru Khmer*). In Indonesia, up to 80% of people consult traditional healers before they consult a medical professional. Traditional healers are also popular in East Timor and Viet Nam. Many Filipino, especially in rural areas, consult *arbolarios* (herbal doctors) who use prayers, herbs, and medicinal plants to heal afflictions (Araneta, 1993). Some also seek help from *manghihilots* who provide *magnetic healing* using prayers and massages similar to acupressure or reflexology (Araneta, 1993; Tan, 2008). Thus, it is important to disaster responders to work with and build capability of local healers (Ito et al., 2012).

STIGMA AND SHAME

Even where mental health programs exist, an important risk factor is whether survivors will actually avail of them. Two important cultural barriers to the provision of postdisaster support is the presence of stigma attached to persons with mental illness and the cultural value of shame. There is much evidence that Asians, in general, are reluctant to seek help from mental health professionals (Hechanova & Waelde, 2017; Matsuoka, Breaux, & Ryujin, 1997; Tuliao, 2014). The reasons are varied. Some are reluctant to open up to people they do not know (Hechanova, Tuliao, Teh, Alianan, & Acosta, 2013). There are also those who simply do not want to burden others (Hechanova & Waelde, 2017). Others are ashamed because seeing mental health professional may mean they are crazy and they do not want to tarnish their dignity or damage their family's reputation (Hechanova & Waelde, 2017). The desire not to lose face is evident in findings that internalized stigma is negatively correlated to the intent to seek professional help (Tuliao, 2014).

THE ROLE OF FAMILY

Given the presence of stigma and shame, survivors in SEA generally prefer seeking help from family and friends (Hechanova et al., 2013) or local healers (Haque, 2010). Fortunately, family relationships are usually protective factors in SEA because families play an essential role in the patients' mental health treatment (Ito et al., 2012).

However, a risk factor is that the lack of knowledge and negative attitudes toward mental illness may prevent people from seeking care (Ito et al., 2012). Thus, it is important for disaster responders to work with survivors' natural counselors – families, community health workers, leaders, religious workers, and healers (Ito et al., 2012; Seekins, 2009; Udomratn, 2008).

COLLECTIVISM

Countries in SEA are described as collectivist and in such cultures, people's identities are interdependent and tied to that of their families, community, and society (Hofstede, 2003). However, collectivism may be a risk factor because even those who are not directly affected may have feelings of self-blame, guilt, and shame when their kin experience trauma (Engelbrecht & Jobson, 2016). A study of Haiyan survivors in the Philippines reported that many expressed feeling guilt and helplessness because they were not able to help their neighbors and friends (Parr, 2015). Alternatively, just as collectivism may influence the impact of disasters, it may also mitigate their impact. Paton et al. (2008) argued that community and institutional factors may influence disaster preparedness in Asia. A study in Indonesia found that the amount of community cooperation influenced the extent to which community members sought information about and prepared for a disaster (Sagala, Okada, & Paton, 2009).

There is also robust evidence that social support cultivates psychological resilience postdisasters (Sasaki et al., 2019). Those in collectivist cultures may have an advantage because they derive their resilience from themselves and their support groups (Haque, 2010). A study among Filipino disaster survivors supported this idea. When asked to reflect on the sources of their strengths, survivors cited the faith in God, family, friends rather than personal characteristics or resources (Hechanova, Waelde, et al., 2015). In addition, social support may actually buffer and increase resistance to the negative effects of a disaster. A study by Sasaki et al. (2019) reported that participants who had emotional and instrumental support prior to a disaster were less likely to report depressive symptoms after a disaster compared to those who did not have support.

One implication of these findings is that interventions that promote social support and interdependent coping efforts may be more important in SEA than interventions that focus only on individual resilience (Hechanova, Ramos, et al., 2015). This idea is corroborated by a study that found that a sense of community promotes the adaptive coping behaviors of first responders (Agbayani, Villaflor, Villaret, & Hechanova, 2019). The authors explained that the sense of community contributed to responders' feeling they are not alone and emboldens them to connect with one another.

All these points suggest the importance of strengthening intact social groups as part of postdisaster interventions. Indeed, studies in the Philippines suggest that group-based interventions are effective in collectivist cultures because they promote healing and empowerment, reduce feelings of shame, decrease alienation and isolation, and discourage passivity and helplessness (Hechanova, Ramos, et al., 2015).

CULTURAL FACTORS THAT MAY INFLUENCE THE DELIVERY OF MHPSS IN SEA

Beyond the vulnerabilities and protective factors, culture can shape how people respond to MHPSS. The Interagency Standing Committee (IASC, 2007) guidelines on the provision of MHPSS in emergency settings emphasize the importance of cultural sensitivity among disaster responders. Beyond understanding the context and language, responders need to understand nuances in social emotions, cognitions, and behaviors to effectively respond to survivors in SEA (Schnyder et al., 2016). Although there is great variability within and between countries in SEA, these nuances in Southeast Asian countries make them different from Western countries and should be considered when providing MHPSS.

SPIRITUALITY

Spirituality shapes attributions about disaster, gives religious significance to hardships and shapes how people cope (Aten, O'Grady, Milstein, Boan, & Schruba, 2014). Religion may also serve as recovery capital because membership in religious groups facilitates both social and instrumental support. There is a diversity of faiths practiced in SEA. Countries such as Malaysia and Indonesia are home to Hinduism which teaches that an individual's mental health depends on a person's actions (karma). Hindus believe that neglecting one's duties toward God and negative behaviors such as cruelty to others, lust, and extortion lead to illnesses. Afflictions are treated using traditional healing practices like prayers, purification, the use of herbal plants, and traditional medicine, yoga and meditation (Haque, 2010).

Buddhism is widely practiced in Thailand, Viet Nam, Cambodia, and Lao and teaches that emotions can also be regulated by the circulation of ch'i (air) and determined by one's food and drink. Buddhist healing approaches typically include acupuncture, qigong, and herbal medicine (Haque, 2010), meditation, reciting the five precepts of Buddhism, and reading Buddhist stories (Fernando, Miller, & Berger, 2010). The blessing and cremation of the dead is also important among Buddhist who believe that spirits may not rest without these rituals. After the 2004 tsunami in Thailand that killed more than 4,000 people, villagers built spirit houses and offered fresh fruit, flowers, and water daily to put spirits to rest (Varley, Isaranuwatchai, & Coyte, 2012).

Islam is also practiced in a number of SEA countries. In Islam, pure thoughts and deeds bring people closer to God and keep them mentally healthy. Traditional Muslim practices use various folk and traditional practices and private prayer are most common form of coping (Ai, Tice, Huang, & Ishisaka, 2005).

Ninety percent of Filipinos belong to Catholic or other Christian denominations. Nakonz and Shik (2009) suggested that Christianity helps Filipinos survive difficult circumstances in three ways. First, hardships are seen as spiritual opportunities to become closer to God or that there is a reason for whatever hardship one is experiencing (i.e., "God never gives you a problem you cannot manage"). Christians seek divine intervention, surrender their hardships to God, and pray for the patience and wisdom to deal with difficulty (Nakonz & Shik, 2009). Tuliao (2014) reported that some Filipino practices prior to Spanish colonization remain. These include the belief that physical and psychological disorders are caused by malevolent spirits and the use of Ayurvedic and Chinese traditional medicine (Lagaya, 2005).

Regardless of spiritual beliefs, there is robust evidence that prayer, belief in God, compassion, and patience play a role in transforming posttraumatic stress experiences into posttraumatic growth (PTG) (Kurniati & Febri, 2014). Beliefs in a higher power and that religion is a good force were correlated to posttraumatic growth among Filipino survivors of Typhoon Haiyan (Nalipay, Bernardo, & Mordeno, 2017).

In addition to the role of spirituality in individual resilience, membership in religious groups also appear to provide recovery support for disaster survivors. A study in Thailand reported that Muslim fishing communities were more cohesive and had better access to support compared to mixed-faith communities (Rigg et al., 2008). In Myanmar, the country's 400,000 monks provide relief and solace to those in need (Seekins, 2009). Among Filipinos, prayer groups, ministries, and religious communities are sources of social support. Help from religious groups, whether in the form of exchange of advice or mobilizing funds to help out those in financial distress, are also important sources of instrumental support (Nakonz & Shik, 2009).

AFFECT AND EXPRESSION

Another critical factor in delivering psychosocial support in SEA is understanding cultural affect, expression, and beliefs. Exposure to trauma leads to negative emotions and appraisals, thought suppression and rumination, and avoidance behaviors that can lead to posttraumatic stress disorder (Dunmore, Clark, & Ehlers, 2001). Western postdisaster therapy tends to focus on talk and expressing emotions. However, some countries in SEA are strongly influenced by Chinese cultures which believe that illness is caused by excessive, unbalanced, or undisciplined emotions (Haque, 2010). In such cultures, survivors may prefer to actively cope by comforting others, helping other survivors, and keeping busy rather than talking about their experiences (Hechanova & Waelde, 2017).

An important finding relevant to affect and expression is that in non-Western cultures, somatic symptoms are emphasized more than the affective component of depression (Tsai & Chentsova-Dutton, 2002). Thus, health professionals need to be aware that survivors may report physical symptoms rather than affective symptoms. When there are no medical explanations for the physical symptoms, more indirect methods such as art and theater activities may be useful to allow survivors to express themselves albeit indirectly (Dueck & Byron, 2012).

POWER DISTANCE

Another cultural nuance in disaster response is the concept of power distance that refers to the extent to which inequality in power is expected in a society (Hofstede, 2003). Countries such as Thailand, Indonesia, Philippines, and Malaysia are classified as high power distance countries (Hofstede, 2003). Power distance may shape expectations of clients of their counselors. A study reported that Asians may expect counselors to be directive and those that are too client-centered and do not provide expert advice may be perceived as incompetent or indifferent (Haque, 2010; Hechanova & Waelde, 2017), though another source has suggested that non-directive counseling may be effective for Thai clients (Tuicomepee, Romano, & Pokaeo, 2012).

Moreover, client interactions with health professionals may be influenced by the extent to which they view the professional as an insider ("one of us") or outsider ("not one of us"). Furthermore, when a health professional is viewed as an outsider, the client may wish to use another person to mediate the communication (Pasco, Morse, & Olson, 2004). One implication of power distance is the need for responders to be relatable to survivors by finding points of commonality such as speaking the same language, discussing that they have undergone similar experiences, or having common family or personal characteristics or experiences. Relatedly, responders may seek to include family members or significant others who may act as an intermediary when power distance may be an issue.

GENDER AND MASCULINITY

Gender is an important consideration in some parts of SEA. For example, Muslim men may expect male counselors and female clients expect female counselors. In many communities, a male chaplain may offer care to a Muslim woman but only when accompanied by the client's male relative (Kirkwood, 2002). There are also some data to suggest gender differences in help seeking. Studies suggest that women are more likely to seek help compared to men in SEA (Gong, Gage, & Tacata, 2003; Tuliao, 2014).

One possible explanation for low help-seeking among males is beliefs about masculinity. Previous work has described Filipino males as less likely to seek help (Valledor-Lukey, 2013). Another study suggested that gender norms prescribe that men should be strong and not show emotional vulnerability (Aguiling-Dalisay et al., 1995). Another study of Filipino male first responders reported that those who believe that males need to endure pain are less likely to cope adaptively and more likely to report vicarious trauma (Agbayani et al., 2019). The authors suggested that this masculinity mindset may lead responders to suppress their negative feelings predisposing them to vicarious trauma.

There are various ways that treatments can be modified to increase engagement. Particularly with SEA culture, treatment access/entry can be adapted by offering services in less stigmatizing locations to reduce shame (Chu & Leino, 2017; Hechanova & Waelde, 2017). Treatment retention/completion and engagement adaptations for SEA culture could include incorporating the family in treatment by providing psychoeducation to reduce stigma, because family plays a central role in SEA individual's mental health (Chu & Leino, 2017; Ito et al., 2012). Finally, treatment delivery could be modified to meet the needs of SEA disaster survivors. Making MHPSS services accessible through disaster relief centers by incorporating the family, spiritual healers, and doctors could potentially improve treatment outcomes for SEA disaster survivors. Treatment satisfaction potentially could improve by incorporating cultural values, beliefs, practices, and examples in MHPSS services.

CONCLUSIONS AND RECOMMENDATIONS

This chapter seeks to add to the sparse literature on MHPSS in SEA by highlighting some common cultural values and traditions, such as the role of spirituality, shame, affect and expression, power distance, gender and masculinity, and collectivism. However, even as we try to describe what SEA countries may have in common, we recognize that here are many variations within cultures and individuals within the SEA region. We acknowledge that there are many other aspects of culture that we have not been able to describe and need to be studied. Much previous work indicates the necessity of cultural sensitivity and understanding local context and culture. Although limited to published work on SEA, we hope that the insights in this chapter can serve as initial starting points that can be validated and explored further by those responding to disasters is SEA.

REFERENCES

Agbayani, B. E., Villaflor, P., Villaret, N., & Hechanova, M. R. (2019). The role of Filipino masculine ideology on the adaptive coping, psychological wellbeing and vicarious trauma of first responders. *International Journal of Culture and Mental Health, 11*, 753–762. https://doi.org/10.1080/17542863.2018.1561736

Aguiling-Dalisay, G., Mendoza, R., Mirafelix, E., Yacat, J., Sto. Domingo, M., & Bambico, F. (1995). *Pagkalalake (masculinity): Men in control? Filipino male views on love, sex and women*. Quezon City: Pambansang Samahan ng Sikolohiyang Pilipino (National Association of Filipino Psychology).

Ai, A., Tice, T., Huang, B., & Ishisaka, A. (2005). Wartime faith-based reactions among traumatized Kosovar and Bosnian refugees in the United States. *Mental Health, Religion & Culture, 8*(4), 291–308.

Araneta, E. G. (1993). Psychiatric care of Filipino Americans. In A. C. Gaw (Ed.), *Culture, ethnicity, and mental illness* (pp. 377–411). Washington, DC: American Psychiatric Press.

Aten, J., O'Grady, K., Milstein, G., Boan, D., & Schruba, A. (2014). Spiritually oriented disaster psychology. *Spirituality in Clinical Practice, 1*, 20–28. https://psycnet.apa.org/doi/10.1037/scp0000008

Bonanno, G. A., Brewin, C. R., Kaniasty, K., & La Greca, A. M. (2010). Weighing the costs of disaster: Consequences, risks and resilience in individuals, families and communities. *Psychological Science in the Public Interest, 11*, 1–49. https://psycnet.apa.org/doi/10.1177/1529100610387086

Chhim, S. (2013). Baksbat (broken courage): A trauma-based cultural syndrome in Cambodia. *Medical Anthropology, 32*, 160–173. https://doi.org/10.1080/01459740.2012.674078

Chu, J., & Leino, A. (2017). Advancement in the maturing science of cultural adaptations of evidence-based interventions. *Journal of Consulting and Clinical Psychology, 85*, 45–57. https://doi.org/10.1037/ccp0000145

Constantine, M. G. (2002). Predictors of satisfaction with counseling: Racial and ethnic minority clients' attitudes toward counseling and ratings of their counselors' general and multicultural counseling competence. *Journal of Counseling Psychology, 49*, 255–263. https://doi.org/10.1037/0022-0167.49.2.255

Dawson, K. S., Joscelyne, A., Meijer, C., Tampubolon, A., Steel, Z., & Bryant, R. (2014). Predictors of chronic posttraumatic response in Muslim children following natural disaster. *Psychological Trauma: Theory, Research, Practice, and Policy, 6*, 580–587. https://psycnet.apa.org/doi/10.1037/a0037140

Dueck, A., & Byron, K. (2012). Community, spiritual traditions, and disasters in Chinese society. *Pastoral Psychology*, *61*, 993–1006. https://doi.org/10.1007/s11089-012-0437-0

Dunmore, E., Clark, D. M., & Ehlers, A. (2001). A prospective investigation of the role of cognitive factors in persistent posttraumatic stress disorder (PTSD) after physical or sexual assault. *Behaviour Research and Therapy*, *39*, 1063–1084. https://doi.org/10.1016/s0005-7967(00)00088-7

Engelbrecht, A., & Jobson, L. (2016). Exploring trauma associated appraisals in trauma survivors from collectivistic cultures. *SpringerPlus*, *5*, 1–11. doi:10.1186/s40064-016-3043-2

Fernando, G. A., Miller, K. E., & Berger, D. E. (2010). Growing pains: The impact of disaster-related and daily stressors on the psychological and psychosocial functioning of youth in Sri Lanka. *Child Development*, *81*, 1192–1210. doi:10.1111/j.1467-8624.2010.01462.x

Gong, F., Gage, S. J. L., & Tacata, L. A. (2003). Helpseeking behavior among Filipino Americans: A cultural analysis of face and language. *Journal of Community Psychology*, *31*, 469–488. doi:10.1002/jcop.10063

Haque, A. (2010). Mental health concepts in Southeast Asia: Diagnostic considerations and treatment implications. *Psychology, Health & Medicine*, *15*, 127–134. doi:10.1080/13548501003615266

Hechanova, M. R. M., Ramos, P. A. P., & Waelde, L. C. (2015). Group-based mindfulness-informed psychological first aid after typhoon Haiyan. *Disaster Prevention and Management: An International Journal*, *24*, 610–618. doi:10.1108/DPM-01-2015-0015

Hechanova, M. R., Tuliao, A. P., Teh, L. A., Alianan, Jr., A. S., & Acosta, A. (2013). Problem severity, technology adoption, and intent to seek online counseling among overseas Filipino workers. *Cyberpsychology, Behavior, and Social Networking*, *16*, 613–617. doi:10.1089/cyber.2012.0648

Hechanova, M. R., & Waelde, L. (2017). The Influence of Culture on Disaster Mental Health and Psychosocial Support Interventions in Southeast Asia. *Mental Health, Religion & Culture, Published online July 5, 2017*. doi:10.1080/13674676 2017.1322048

Hechanova, M. R. M., Waelde, L. C., Docena, P. S., Alampay, L. P., Alianan, A. S., Flores, M. J. B., … Melgar, M. I. E. (2015). The development and initial evaluation of Katatagan: A resilience intervention for Filipino disaster survivors. *Philippine Journal of Psychology*, *48*, 105–131. Retrieved from https://mhpss.net/resource/the-development-and-initial-evaluation-of-katatagan-a-resilience-intervention-for-filipino-disaster-survivors/

Hofstede, G. (2003). *Culture's consequences: Comparing values, behaviors, institutions and organizations across nations*. Thousand Oaks, CA: Sage.

Interagency Standing Committee. (2007). IASC guidelines on mental health and psychosocial support in emergency settings. Geneva: IASC.

Ito, H., Setoya, Y., & Suzuki, Y. (2012). Lessons learned in developing community mental health care in East and Southeast Asia. *World Psychiatry*, *11*, 186–190.

Kirkwood, N. A. (2002). *Pastoral care to Muslims: Building bridges*. Binghamtom, NY: The Hawthorn Pastoral Press.

Lagaya, A. T. (2005). Republic of the Philippines. In G. Bodeker, C. K. Ong, C. Grundy, G. Burford, & K. Shein (Eds.), *WHO global atlas of traditional, complementary, and alternative medicine* (pp. 199–204). Kobe: World Health Organization Centre for Health Development.

Lehman, D. R., Chiu, C., & Schaller, M. (2004). Psychology and culture. *Annual Review of Psychology*, *55*, 689–714. https://doi.org/10.1146/annurev.psych.55.090902.141927

Matsuoka, J. K., Breaux, C., & Ryujin, D. H. (1997). National utilization of mental health services by Asian Americans/pacific islanders. *Journal of Community Psychology*, *25*, 141–145. https://doi.org/10.1002/(SICI)1520-6629(199703)25:2%3C141::AID-JCOP3%3E3.0.CO;2-0

Nakonz, J., & Shik, A. W. Y. (2009). And all your problems are gone: Religious coping strategies among Philippine migrant workers in Hong Kong. *Mental Health, Religion & Culture*, *12*, 25–38. https://doi.org/10.1080/13674670802105252

Nalipay, M. J., Bernardo, A. B. I., & Mordeno, I. G. (2017). Posttraumatic growth in survivors of a natural disaster: The role of social axioms of religiosity, reward for application and social cynicism. *Journal of Positive Psychology*, *12*, 342–353. https://doi.org/10.1080/17439760.2016.1187199

Parr, R. (2015). The use of expressive arts in making sense of traumatic experiences. *Philippine Journal of Psychology*, *48*, 133–159. Retrieved from https://www.pap.org.ph/includes/view/default/uploads/pjp-journals/PJP1502_Final_6Parr.pdf

Pasco, A. C. Y., Morse, J. M., & Olson, J. K. (2004). Cross-cultural relationships between nurses and Filipino Canadian patients. *Journal of Nursing Scholarship*, *36*(3), 239–246.

Paton, D., Gregg, C. E., Houghton, B. F., Lachman, R., Lachman, J., Johnston, D. M., & Wongbusarakum, S. (2008). The impact of the 2004 tsunami on coastal Thai communities: Assessing adaptive capacity. *Disasters*, *32*, 106–119. https://doi.org/10.1111/j.1467-7717.2007.01029.x

Rigg, J., Grundy-Warr, C., Law, L., & Tan-Mullins, M. (2008). Grounding a natural disaster: Thailand and the 2004 tsunami. *Asia Pacific Viewpoint*, *49*, 137–154. https://doi.org/10.1111/j.1467-8373.2008.00366.x

Rolfsness, E. S., & Idsoe, T. (2011). School-based intervention programs for PTSD symptoms: A review and meta-analysis. *Journal of Traumatic Stress*, *24*, 155–165. doi:10.1002/jts.20622

Sagala, S., Okada, N., & Paton, D. (2009). Predictors of intention to prepare for volcanic risks in Mt Merapi, Indonesia. *Journal of Pacific Rim Psychology*, *3*, 47–54. https://doi.org/10.1375/prp.3.2.47

Samphantharak, K. (2014). Natural disasters and the economy: Some recent experiences from Southeast Asia. *Asian-Pacific Economic Literature*, *28*, 33–51. https://doi.org/10.1111/apel.12075

Sasaki, Y., Aida, J., Tsuji, T., Koyama, S., Tsuboya, T., Saito, T., … Kawachi, I. (2019). Pre-disaster social support is protective for onset of post-disaster depression: Prospective study from the Great East Japan Earthquake and Tsunami. *Nature*, *9*, 19427. https://doi.org/10.1038/s41598-019-55953-7

Schnyder, U., Bryant, R., Ehiers, A., Foa, E., Hasan, A., Mwiti, G., … Yule, W. (2016). Culture-sensitive psychotraumatology. *European Journal of Psychotraumatology*, *7*, 31179. https://doi.org/10.3402/ejpt.v7.31179

Seekins, D. M. (2009). State, society and natural disaster: Cyclone Nargis in Myanmar (Burma). *Asian Journal of Social Science*, *37*, 717–737. doi:10.1163/156848409X12474536440500

Subandi, M. A., Achmad, T., Kurniati, H., & Febri, R. (2014). Spirituality, gratitude, hope and post-traumatic growth among the survivors of the 2010 eruption of Mount Merapi in Java, Indonesia. *Australasian Journal of Disaster and Trauma Studies*. *18*(1), 19–26.

Tsai, J. L., & Chentsova-Dutton, Y. (2002). Understanding depression across cultures. In I. Gotlib & C. Hammen (Eds.), *Handbook of depression* (pp. 467–491). New York, NY: Guilford Press.

Tuicomepee, A., Romano, J. L., & Pokaeo, S. (2012). Counseling in Thailand: Development from a Buddhist perspective. *Journal of Counseling and Development*, *90*, 357–361. https://doi.org/10.1002/j.1556-6676.2012.00044.x

Tuliao, A. (2014). Mental health help seeking among Filipinos: A review of the Literature. *Asia Pacific Journal of Counseling and Psychotherapy*, *5*, 124–136.

Udomratn, P. (2008). Mental health and the psychosocial consequences of natural disasters in Asia. *International Review of Psychiatry*, *20*, 441–444. doi:10.1080/09540260802397487

United Nations International Strategy for Disaster Reduction. (2010). Protecting Development Gains. Reducing Disaster Vulnerability and Resilience in Asia and the Pacific. The Asia Pacific Disaster Report 2010. Retrieved from https://www.unescap.org/sites/default/files/publications/Asia-Pacific-Disaster-Report-2010.pdf

Valledor-Lukey, V. V. (2013). Pagkababae at pagkalalake (femininity and masculinity): Developing a Filipino Gender Trait Inventory and predicting self-esteem and sexism. Dissertation Abstracts International Section A, 74.

van Griensven, F., Chakkraband, M. L. S., Thienkrua, W., Pengjuntr, W., Cardozo, B. L., Tantipiwatanaskul, P., … Tappero, J. W. (2006). Mental health problems among adults in tsunami-affected areas in Southern Thailand. Journal of the American Medical Association, *296*, 537–548. doi:10.1001/jama.296.5.537

Varley, E., Isaranuwatchai, W., & Coyte, P. C. (2012). Ocean waves and roadside spirits: Thai health service providers' post-tsunami psychosocial health. *Disasters*, *36*, 656–675. doi:10.1111/j.1467-7717.2012.01274.x

Warsini, S., Buettner, P., Mills, J., West, C., & Usher, K. (2014). The psychosocial impact of the environmental damage caused by the MT Merapi eruption on survivors in Indonesia. *EcoHealth*, *11*, 491–501. doi:10.1007/s10393-014-0937-8

PART II

LEADERSHIP AND ORGANIZATION FOR MENTAL HEALTH AND PSYCHOSOCIAL SUPPORT

CHAPTER 2

EDUCATION AND DISASTER VULNERABILITY IN SOUTHEAST ASIA: EVIDENCE AND POLICY IMPLICATIONS

Roman Hoffmann and Daniela Blecha

ABSTRACT

This chapter summarizes the growing theoretical and empirical literature on the role of education in disaster risk reduction with a focus on Southeast Asia. Education and learning can take place in different environments in more or less formalized ways. They can influence disaster vulnerability in direct and indirect ways. Directly, through education and learning, individuals acquire knowledge, abilities, skills and perceptions that allow them to effectively prepare for and cope with the consequences of disaster shocks. Indirectly, education gives individuals and households access to material, informational and social resources which can help reduce disaster vulnerability. This chapter highlights central concepts and terminologies and discusses the different theoretical mechanisms through which education can support disaster risk reduction. Supportive empirical evidence is presented and discussed with a particular focus on the role of inclusiveness in education and challenges in achieving universal access to high-quality education. Based on a situation analysis and best practice cases, policy implications are derived that can inform the design and implementation of education and learning-based disaster risk reduction efforts in the region.

Keywords: Education; disaster vulnerability; disaster risk reduction; Southeast Asia; review; policy implications

Resistance, Resilience, and Recovery from Disasters: Perspectives from Southeast Asia
Community, Environment and Disaster Risk Management, Volume 21, 17–37

ISSN: 2040-7262/doi:10.1108/S2040-726220200000021002

Who is vulnerable to disasters and other environmental hazards and to what extent are fundamental questions in vulnerability reduction efforts? This chapter focuses on differential vulnerabilities to natural hazards in Southeast Asia and reviews evidence of the role of formal, non-formal, and informal education in influencing households' and communities' preparedness against and ability to cope with hazards and shocks. Education has been shown to be an important enabler of sustainable development in different contexts and with relevance for various development outcomes (Bengtsson, Barakat, & Muttarak, 2018). It is not only essential to poverty alleviation and economic growth but can also enhance societies' adaptive capacity vis-à-vis future climate change (Lutz, Muttarak, & Striessnig, 2014; Muttarak & Lutz, 2014). As such, education, including both school-based and less formalized approaches, can also make an important contribution to reducing vulnerability to natural hazards.

The important role of education in disaster risk reduction and mitigation has been recognized in the Sendai Framework for Disaster Risk Reduction, the first major United Nations (UN) agreement of the Sustainable Development Agenda. The Sendai Framework highlights the need for population-wide resilience building and the enhancement of disaster preparedness for effective response and fast recovery, rehabilitation, and reconstruction. To achieve these aims, the framework underlines the importance:

> to promote national strategies to strengthen public education and awareness in disaster risk reduction, including disaster risk information and knowledge, through campaigns, social media and community mobilization, taking into account specific audiences and their needs. (United Nations Office for Disaster Risk Reduction (UNISDR), 2015, p. 14)

Despite improvements in national disaster risk reduction efforts in the Southeast Asian region, especially after major disaster events such as the 2004 Indian Ocean Tsunami (Birkmann, 2008) or the 2013 Typhoon Haiyan in the Philippines (Lagmay et al., 2015), further steps have to be taken to prevent loss of life and major damages in the future. Often, disaster risk reduction measures undertaken at the national level alone are not enough to protect households and communities from the devastating impacts of natural hazards. Not only are public efforts challenged with addressing local needs and conditions, it also takes some time for local government and disaster-relief organizations to mobilize resources to an affected area. In addition to national initiatives, increased local efforts are needed, involving the communities and individual households, to help withstand future shocks (Hoffmann & Muttarak, 2017).

This chapter places a particular focus on the vulnerability to natural hazards and calamities. These are rapid-onset, extreme events that result from natural (e.g., hydrological, meteorological, geological or climatological) processes. Examples include floods, typhoons, volcanic eruptions, tsunamis or earthquakes. Due to their limited predictability, natural hazards can cause severe loss of life and major damages to property. In order to prevent disastrous consequences, adequate preparation and the undertaking of precautionary measures is a key.

In addition to community efforts, household preparedness measures such as stockpiling of food and water, having a first aid kit in the home, or having a family evacuation plan can help mitigating the harmful consequences of disasters (Shreve & Kelman, 2014; Van der Keur et al., 2016). Despite the importance of individual preparedness, several studies report relatively low levels of preparedness even in disaster-prone areas (Kohn et al., 2012). In this context, education can help raising awareness among households and promote the adoption of precautionary measures. It can furthermore support the provision of critical material and immaterial resources, required to adequately prepare against and adapt to natural hazards. Education can hence be a key element in disaster risk mitigation and help reducing vulnerabilities to disaster and other environmental hazards both on a household and community level.

The remainder of this chapter is structured as follows. The next section introduces central concepts related to education and demographic differential vulnerabilities to natural hazards, followed by a review of the theoretical and empirical literature. In particular, the review focuses on potential channels through which education can have an impact on related empirical evidence. The next section presents a situation analysis for the case of Southeast Asia and highlights major challenges in the education systems in the regions that are of relevance for disaster risk reduction efforts. Based on the findings, we showcase best practices of education policies and interventions that have been successfully implemented in the region and derive policy lessons for national and local disaster risk reduction efforts. The final section of the chapter concludes with some final remarks and an outlook on future research. Parts of this chapter were also published in the journal Sustainability (Hoffmann & Blecha, 2020, special issue "Education and Sustainable Development," ISSN 2071-1050).

CENTRAL CONCEPTS AND TERMINOLOGY

We follow a broad definition of education as the process of facilitating the learning of knowledge, skills/competencies, values, beliefs and habits. People are constantly learning in different environments. Education can take place in formalized institutions as well as in non-formal or informal settings. Formal education or learning refers to classroom-based education normally delivered in a systematic way by trained teachers in a structured environment such as a school, college or university. This form of education is intentional and has specific pre-defined learning outcomes, which are transparently communicated to the learners.

Education, however, is not restricted to learning in the public education system but can take place in different environments outside the classroom. As we will show later, these often less institutionalized forms of information and knowledge dissemination can play a crucial role in reducing vulnerabilities to disasters. Organized and structured education activities that takes place outside the established formal school system are referred to as non-formal education (Singh, 2016; Werquin, 2010). These include community education or alternative learning programs, such as community-based disaster trainings and drills. Informal education, on the other hand, refers to learning through exchange with others and the enlargement of experiences. This less structured form of education includes social learning from peers or family members. In particular, knowledge, skills, values and beliefs obtained during childhood through parents and other close family members can be a crucial factor in influencing behavior later in life (Marsick & Watkins, 2002; Richardson & Wolfe, 2001).

Importantly, learned knowledge and skills are not always explicit to the learner, but can be tacit. Explicit knowledge can be clearly identified, accessed and verbalized. Tacit knowledge, on the other hand, is implicit (Linde, 2001; Nonaka & von Krogh, 2009). It can be defined as knowledge, skills, ideas and experiences that people have but which they may not be consciously aware of and cannot necessarily be easily expressed or explained to others. This involves knowledge about how to perform certain activities, such as riding a bike or tying a shoe, or how to behave and react in certain situations. Tacit knowledge is typically learned through interactions with others and is thus closely connected to informal learning. Acquiring it requires practice and extensive social exchange. Forms of internalized knowledge and behavioral skills are of great importance for disaster prevention. Especially under stress or in life-threatening situations, humans tend to follow internalized heuristics, intuitions, and implicit knowledge and skills. These can be decisive for whether people take the right decisions and can mean the difference between life and death (Weichselgartner & Pigeon, 2015).

Disaster education or disaster resilience education, which is any form of learning about risks of environmental hazards and protective measures, should address both explicit as well as tacit knowledge and skills. In the recent years, the field of disaster education has grown substantially, including new ways to prepare the public for disasters that go beyond school-based initiatives and public information campaigns, but also include more non-formal ways of learning, such as family and community learning activities and adult and senior citizen education. New technological advancements, such as social media and disaster apps, open further pedagogical space for learning and the dissemination of information and knowledge.

Vulnerability to environmental hazards is a function of a household's or community's susceptibility, sensitivity and adaptive capacity (Intergovernmental Panel on Climate Change (IPCC), 2014). Susceptibility refers to the presence of people in places that are at risk of being adversely affected, while sensitivity relates to the degree to which a population group is negatively affected once a shock occurs. Adaptive capacity is the ability to cope with and adjust to consequences of a natural hazard. This chapter considers all three aspects of vulnerability. Importantly, within societies, different population groups are characterized by a differential vulnerability, that is, environmental hazards do not affect population groups equally. The degree to which a population group is vulnerable depends on social, economic, cultural and political factors, which are often closely interlinked with each other. Socioeconomic inequalities play an important role with marginalized groups, women, young children and older people, as well as people with health problems or disabilities being particularly vulnerable to natural hazards (Muttarak, Lutz, & Jiang, 2015; Zagheni, Muttarak, & Striessnig, 2015).

In this regard, resilience building on household and community levels helps reducing vulnerabilities in advance. It is an important part of any disaster risk mitigation strategy. The term resilience refers to the household's or community's abilities and resources to mitigate, adequately prepare for and cope with disruptive changes and events that may occur unforeseen. Various factors contribute to resilience, including not only economic, but also social, demographic and institutional factors. Among these, education and the dissemination of knowledge and information can also effectively contribute to resilience building efforts (Paton & Johnston, 2001).

THEORETICAL FRAMEWORK: UNDERSTANDING THE UNDERLYING MECHANISMS

Education can influence disaster vulnerability through various channels, which can be distinguished in direct and indirect effects. Fig. 1 illustrates the different pathways in a schematic way. Importantly, here, we focus on all forms of education, be it formal, non-formal, or informal, which can take place in different environments. Primarily, these are expected to affect disaster vulnerability by positively influencing household and community resources (material, informational and social), awareness and beliefs and risk perceptions, which in turn affect the susceptibility, sensitivity and adaptive capacity as main vulnerability features.

First, education gives individuals access to knowledge and information about threats as well as about ways to adequately take measures against them. In particular, when it comes to preparing against a disaster, having comprehensive information is crucial. This includes information about the disaster risks in a community (e.g., where and when to expect a disaster), the adequacy

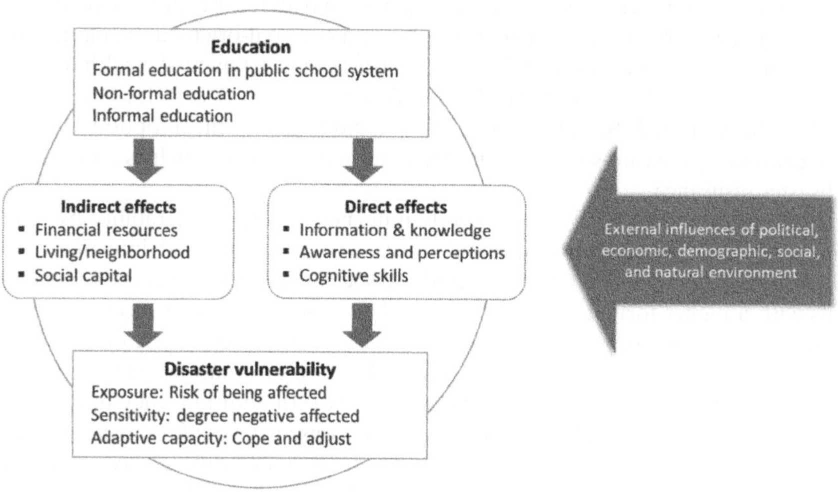

Fig. 1. Conceptual Relationship Between Education and Disaster Vulnerability.

of precautionary measures (e.g., stockpiling of food and emergency supplies), warning systems (e.g., about from where to receive further information), emergency plans and management (e.g., institutions providing support during a disaster), how to respond and behave during a disaster (e.g., where to look for shelter) and ways to cope with a disaster in the aftermath (e.g., when to return to the disaster affected area). These are not only important for households, but also, for entire communities, who require knowledge and information for informed decision-making and coordinated actions (Weichselgartner & Pigeon, 2015).

At the same time, through education, individuals acquire skills and competencies that can help reducing their vulnerability to external shocks. For formal schooling, there is established evidence showing how schooling can positively affect cognitive skills, including complex cognitive reasoning such as categorization, logical deduction and the solving of complex problems (Blair, Gamson, Thorne, & Baker, 2005; Ceci, 1991; Eslinger et al., 2009; Lee, 2010; Quartz & Sejnowski, 1997). Such cognitive skills can be helpful in assessing the risks related to disaster threats and in finding adequate responses (De Bruin, Parker, & Fischhoff, 2007; Peters et al., 2006).

Literacy skills obtained at school can also reduce disaster vulnerability. Literacy does not only refer to the ability to read and write, but also to understand information, such as warnings or recommendations. Functional literacy refers to the ability of a person to:

> engage in all those activities in which literacy is required for effective functioning of his or her group and community and also for enabling him or her to continue to use reading, writing and calculation for his or her own and the community's development. (United Nations Educational, Scientific and Cultural Organization (UNESCO), 2019a)

Education could also directly benefit disaster preparation efforts by changing risk and time preferences as well as the capacity to plan for the future. Recent evidence suggests that more educated people act more goal-oriented, being able to better allocate resources and making more investments in financial, health or education for their future (Chew, Heckman, Yi, Zhang, & Zhong, 2010; Grossman, 2006; Oreopoulos & Salvanes, 2011). This could influence the adoption of such precautionary measures which require long term investments, such as purchasing disaster insurance.

In addition, education can indirectly influence disaster vulnerability by having a positive impact on household's access to economic and informational resources, their settlement decision, and the social capital. On average, individuals with a higher formal education and higher education degrees receive greater earnings and have more possibilities to diversify their income in case one of their income sources can no longer be sustained because of a natural hazard (Card, 1999; Heckman, Humphries, & Veramendi, 2018). In addition, they have more resources at their disposal to buffer the negative consequences of a shock. Greater economic resources also enable the better educated to undertake costly preparedness actions, such as structural adjustments in their housing or the purchase of an insurance against disaster risks. They can also use their resources to relocate from risk zones, such as flooding areas, making them less likely to be directly exposed to and affected by the negative effects of disasters.

Numerous studies have shown that education is positively related with better access to informational resources, that is, more sources and types of information (Cotten & Gupta, 2004; Neuenschwander, Abbott, & Mobley, 2012; Wen, Rissel, Baur, Lee, & Simpson, 2011). This is partly due to a greater knowledge about where to obtain information from and better resources and skills to receive information, such as better access to information and communication technologies. For instance, it has been shown that the level of formal education is highly correlated with access to weather forecasts and early warnings, allowing better educated individuals to faster respond and prepare against hazards (Rodriguez et al., 2007; Xiao & McCright, 2007).

Another important channel through which education may increase preparedness activities is via social capital, that is, resources that can be accessed through the social networks of peers and kin (Huang, Van den Brink, & Groot, 2009; Lake & Huckfeldt, 1998). Both formalized and less formalized forms of education can help people understand the importance of social ties in disaster preparation and adaptation. Individuals who participate regularly in social activities can benefit from an exchange of useful information and warnings. Indeed, there is evidence that social participation, strong family and community networks and high level of trust are positively associated with preparedness behavior (Kirschenbaum, 2006; Solberg, Rossetto, & Joffe, 2010; Witvorapong, Muttarak, & Pothisiri, 2015).

Importantly, the aforementioned factors interact with each other in a complementary way. Having material resources alone is often not enough. Individuals can benefit from material resources only with a thorough understanding of how to put the resources to the greatest use. Likewise, without access to financial or social resources, it is impossible to adequately prepare against a disaster shock, even with knowledge and awareness of its potentially harmful consequences. These multidimensional and complex interactions between different factors make it difficult to tackle vulnerabilities without concomitantly addressing the different underlying factors.

EMPIRICAL EVIDENCE: EDUCATION EFFECTS ON DISASTER VULNERABILITY

There is a growing empirical literature addressing the question how education affects disaster vulnerability and how it can be used to strengthen resilience of households and communities. This section reviews selected findings from the literature related to the effects of different forms of education on vulnerability outcomes: Susceptibility, sensitivity and adaptive capacity. Where possible, we highlight, based on the conceptual model developed in the previous section, the channels through which education and learning may have an impact and why. In the literature, there is a clear focus on the effects of formalized education on vulnerability outcomes, which is also reflected in this review.

Commonly, it is households with lower socioeconomic status and lower education level who reside in high-risk areas, whereas higher education groups are more likely to live in low-risk areas. The better educated have been shown to be also more likely to relocate if they find themselves in a risky location. This relocation can result from different risk perceptions, a greater awareness of the potential hazards, or greater resources enabling them to move (KC, 2013; Muttarak & Lutz, 2014; Wamsler, Brink, & Rentala, 2012). Also for short-term evacuations, Sharma, Patwardhan, and Patt (2013) show that better educated persons react most strongly to cyclone warnings and are the ones most likely to evacuate after receiving a warning.

During the pre-disaster phase, disaster mitigation measures and the undertaking of precautionary actions are important. Numerous studies report that education, be it general education or specific disaster trainings, increase individual and household preparedness, including preparedness for earthquakes (Russell, Goltz, & Bourque, 1995), hurricanes (Baker, Leon, Smith Greenaway, Collins, & Movit, 2011; Norris, Smith, & Kaniasty, 1999; Reininger et al., 2013), floods (Lave & Lave, 1991; Thieken, Kreibich, Müller, & Merz, 2007), tsunami (Muttarak & Pothisiri, 2013), as well as general emergency preparedness (Al-Rousan, Rubenstein, & Wallace, 2014; Smith & Notaro, 2009). Similar findings are reported not only on individual but also on aggregate level in country comparisons (Pichler & Striessnig, 2013).

Hoffmann and Muttarak (2017) and Muttarak and Pothisiri (2013) showed for the Philippines and Thailand that individual and household education levels are positively correlated with the probability of a household having undertaken precautionary measures and being prepared against disasters. Interestingly, the authors show that households also learn from previous disasters. If a household was affected by a disaster in the past, its preparedness levels increases substantially. Similar evidence has been reported for households in other countries, which have experienced floods, earthquakes, hurricanes and wildfires (Kohn et al., 2012). Although learning through experience is common, it is not an ideal way to promote household adoption of precautionary measures. Instead, educational initiatives can fill the gap leading to higher degrees of preparation, even if personal experiences with disasters are missing.

In the aftermath of a disaster, education can help people to cope with the events and to adapt to the changed environment. Using longitudinal representative data from the island of Sumatra, Indonesia, before and after the 2004 Indian Ocean tsunami, Frankenberg, Sikoki, Sumantri, Suriastini, and Thomas (2013) showed a higher survival rate for better educated males, but not females, suggesting important interactions with the local social and cultural context. Although the authors do not find differences in the level of posttraumatic stress, the higher educated were better able to cope with the disaster in the long term. For instance, they were less likely to live in a camp or temporary housing, but rather stayed with family and friends, revealing the important role of social networks for recovery after a disaster event (Guarnacci, 2016; Marín, Bodin, Gelcich, & Crona, 2015). At the same time, the better educated were more effective in smoothing and stabilizing their consumption after the disaster shock, that is, they were exposed to less fluctuations and instability, revealing a greater economic resilience to hazards. In another study about the 2004 disaster event in Indonesia, Irmansyah, Dharmono, Maramis, and Minas (2010) found lower levels of anxiety, affective disorders and posttraumatic stress among better educated survivors of the disaster, indicating a higher resilience in terms of psychosocial wellbeing. Similar findings have been reported for Thailand (Garbero & Muttarak, 2013).

Aside of general formal education, various studies have reported positive effects of non-formal education initiatives, including community-based disaster trainings and drills. These have been shown to be effective in building local capacities and in raising awareness and resilience of communities (Allen, 2006; Karanci, Aksit, & Dirik, 2006). Muttarak and Pothisiri (2013), who conducted their study in risk areas in coastal Thailand, reported that households who participated in disaster trainings were more likely to undertake precautionary measures. The trainings were most effective for individuals with high education, suggesting interactions between different education forms. Despite the potential of directed/steered disaster risk trainings, studies also point to challenges and barriers in their implementation, such as inadequate communication, lack of integration with the communities, and non-inclusivity, which might serve as an explanation for why some studies do not find a benefit.

Education is not only of relevance for household vulnerability, but also for the vulnerability of entire communities. In fact, it has been found that living in a community with a higher level of education significantly increases preparedness actions (Muttarak & Pothisiri, 2013). This suggests there is knowledge spillovers and skill exchange amongst community members. Hoffmann and Muttarak (2017) suggested that non-formal or social learning in the affected communities play an important role. Households in a community can exchange their experiences – possibly after having experienced a disaster event together – and undertake effective joint actions to prepare against future disaster events. Such complementarities in the communities can generate positive spill-overs and amplify the impact of education for the learners.

SITUATION ANALYSIS: EDUCATION AND RISK REDUCTION IN SOUTHEAST ASIA

The foregoing review indicates that both school-based and other forms of education matter for reducing disaster vulnerabilities in high-risk environments. This section describes the situation in Southeast Asia and analyzes to what extent insights from the education and learning literature have found entry in national disaster risk reduction efforts. In particular, we discuss challenges related to providing inclusive access to education for all.

The formal education system and institutions in Southeast Asia differ from country to country and are strongly shaped by the local context. The heterogeneities result in different needs and challenges which must be considered. Fig. 2 shows the net-enrollment rates (NER), that is, the total number of enrolled students in a relevant age group divided by the total population in that age group,

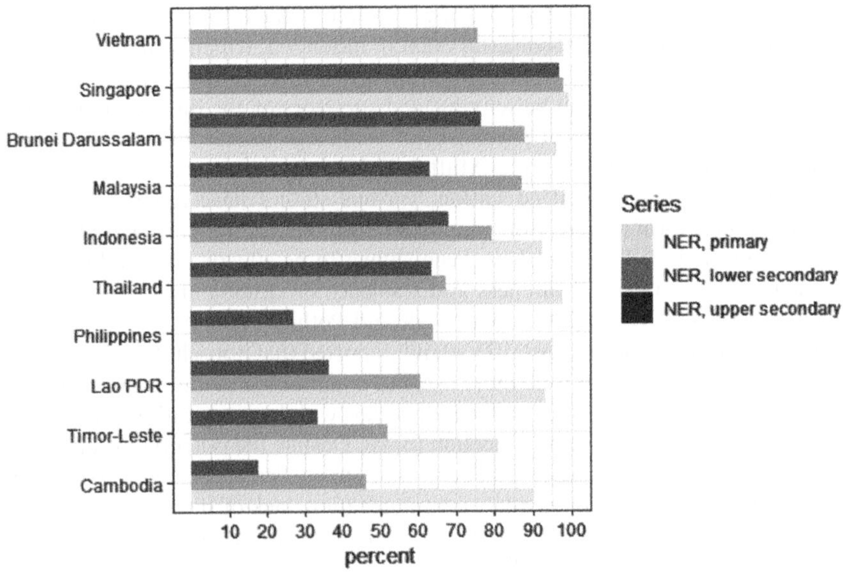

Fig. 2. Net-enrollment Rates in Different Grades for Countries in Southeast Asia. *Source*: World Bank Education Statistics (EdStats) 2019. Percentages Calculated Based on Data From Last Available Year.

in different grades for countries in Southeast Asia (World Bank, 2019). While net enrollment in primary grades is generally high across the region, substantial differences in the NET for lower and upper secondary education can be observed with poorer countries showing on average lower enrollment levels.

Importantly, educational resources and access to basic education are not equally distributed in the population, but are strongly influenced by social factors and circumstances, such as wealth, ethnicity and location. Typically, people have little control over these factors, even though they play an important role in shaping their opportunities. Educational inequality refers to the unequal distribution of and access to academic resources. Poverty and marginalization are two of the major causes of educational inequalities, partly resulting from disadvantages in the education systems.

Fig. 3 illustrates educational inequalities for selected countries in Eastern and Southeast Asia. It shows the lower secondary education completion rates for different wealth quintiles (poorest vs richest population groups). The black diamond shapes reflect the average completion rates. The length of the black lines indicates the amount of inequality in the countries. The graph reveals substantial educational inequalities in some Southeast Asian countries, such as Lao PDR, Cambodia or Myanmar, where less than 25% of the poorest population group have completed lower secondary education.

Educational inequalities have important implications for disaster risk reduction. As they are more likely to reside in high-risk areas, impoverished and

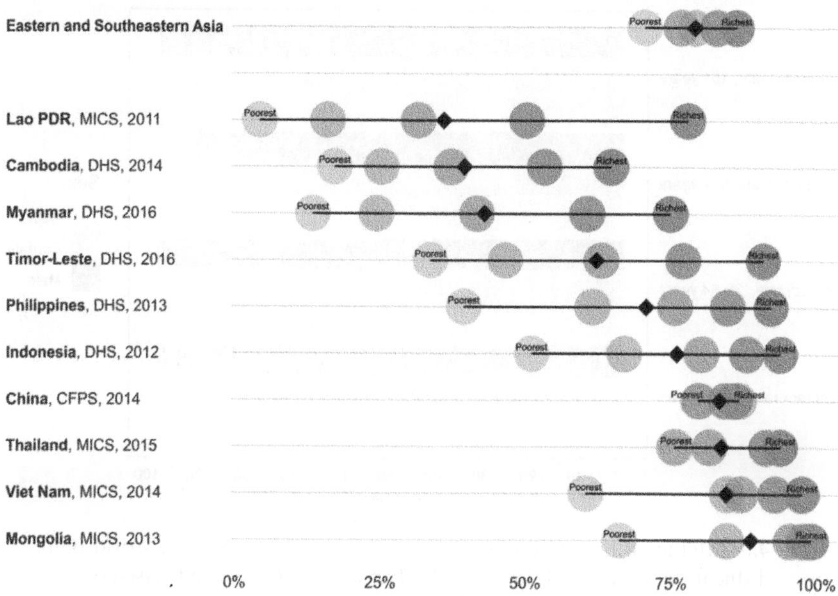

Fig. 3. Educational Inequalities in Eastern and Southeast Asia. Graph Shows Lower Secondary Education Completion Rates for Different Wealth Quintiles. Diamond Shapes Reflect the Average Completion Rates. Data Based on Household Surveys: Multiple Indicator Cluster Surveys (MICS), Demographic and Health Survey (DHS), China Family Panel Survey (CFPS). The Year Refers to the Year of Data Collection. *Source*: UNESCO, World Inequality Database on Education.

marginalized communities are on average more exposed to environmental hazards and lack material resources to adequately prepare against hazardous events. Their increased exposure and limited resources, together with their deprivation from education, lead to a substantial increase in disaster vulnerability among these population groups. In this regard, especially rural areas often face a disadvantage. Rural populations, who have commonly worse access to education, are on average not only more exposed to natural hazards, they also depend more on the environment in their livelihoods.

Providing access to universal and high-quality education for all population groups can generate important spill-over effects for disaster risk reduction. In addition to providing universal access to formal education, non-formal education, such as disaster risk reduction trainings, can also help counteracting existing gaps and serve as a short-term remedy addressing some of the challenges and risks faced by marginalized population groups.

In addition to social factors, gender and age are important demographic factors influencing educational attainment. Fig. 4 shows literacy rates in Southeast Asia by age groups and sex. Although there are very high literacy rates for both males and females in the age group 15–64, the rate is with 80% substantially lower

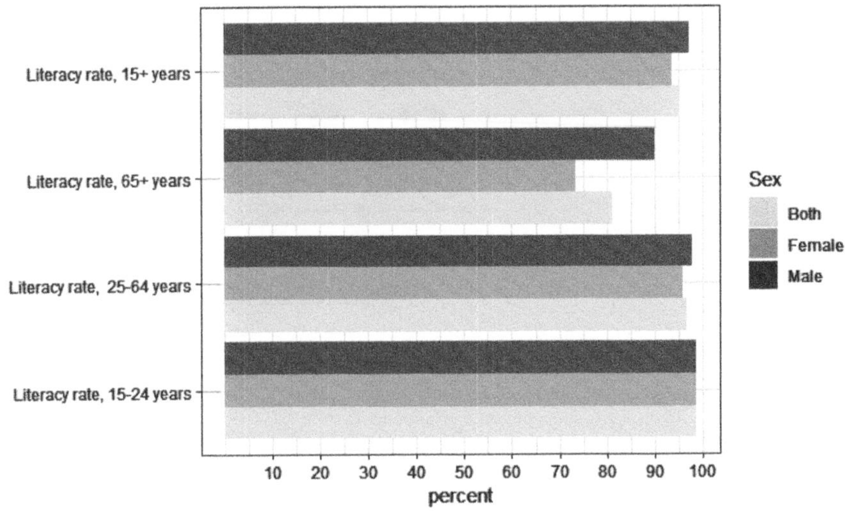

Fig. 4. Literacy Rates in Southeast Asia by Age and Sex. *Source*: World Bank Education Statistics (EdStats) 2019. Percentages Calculated Based on Data from Last Available Year.

in the oldest age category of 65+ years. In this group, women have a disproportionally lower literacy compared to men, making them particularly vulnerable to environmental hazards. Increased levels of vulnerability among the elderly in general, and older women in particular, have been shown in different studies (Kohn et al., 2012). Such vulnerabilities result from complex interactions between environmental, social and demographic factors and can be reinforced through multiple sources.

Achieving universal education and fighting educational inequalities was made a top priority in the recently adopted Sustainable Development Agenda (UN, 2015). In addition, countries in Southeast Asia have embarked on reform processes to ensure inclusive and equitable quality education and promote lifelong-learning opportunities for all. These reforms have resulted in improvements in access to universal education and educational attainments across all social groups. At the same time, major investments in school infrastructures and teachers led to higher education quality and better learning outcomes.

The role of education and learning has also been increasingly recognized in countries' disaster risk reduction efforts and strategies. Disaster trainings and drills have been successfully integrated in school curricula, for example in the Philippines, Indonesia and Thailand. Also, outside school, children and youth are increasingly involved in community efforts and political processes related to disaster risk reduction. Special disaster training program has been implemented that address vulnerable groups by taking into account the manifold interactions that create disaster vulnerabilities. For instance, special trainings have been developed to address senior citizens following a lifelong-learning approach and considering their particular needs and challenges.

Together with infrastructural measures on different levels, these interventions represent important steps to involve populations in disaster risk reduction efforts through education and learning. Despite the progresses, major challenges remain related to the inclusiveness of education and to reducing disaster vulnerabilities along all dimensions. The following section presents best practice policy cases and highlight policy lessons of relevance for the Southeast Asian region.

BEST PRACTICE EXAMPLES

Formal learning pathways are important enablers for disaster resilience education. In the classroom setting, disaster risk reduction courses can either be offered as stand-alone courses or be integrated into the existing school curriculum. For the latter, it is necessary to undertake an audit of the existing curriculum and to appropriately define the entry points in the course of the curriculum adoption cycle for all subjects and age levels.

A promising example of formal disaster education is the ASEAN's *Safe Schools Initiative* (ASSI), which was launched in 2012 and promotes the inclusion of disaster risk reduction in national curricula across eight South-East Asian countries, including Cambodia, Indonesia, Lao PDR, Malaysia, Myanmar, Philippines, Thailand and Vietnam (ASEAN, 2018). In partnership with ASEAN Member States and several non-governmental organizations such as Plan International, Save the Children, World Vision and Mercy, and funding from the European Union and Australian Aid, the initiative aims at building up disaster risk reduction and resilience of school communities through education. Pillar three of its framework, which sets the guidelines for any practical intervention, focuses on "risk reduction and resilience education" and is aligned with disaster management and education sector plans at national, sub-national and local levels.

A flagship program of the Safe Schools Initiative was launched in Malaysia in 2018, consisting of hands-on safety preparedness lessons for students and teachers to equip them with a basic understanding of the concepts of disaster risk reduction plans as well as how to prepare for and act in the event of an emergency. While it intends to reach 1,000 students and 100 teachers in 20 schools around the country, the material is only available in English and Bahasa Malaysia languages, which led to the exclusion of significant Mandarin speaking population groups in some provinces of Malaysia. In Thailand, the initiative has succeeded in training more than 5,000 pupils from 40 schools in disaster risk reduction, including the conduction of practical safe school exercises. According to involved teachers and civil society organizations, students successfully learnt how to save drowning people as well as themselves.

The implementation of the initiative in the countries was confronted with various challenges, such as a limited willingness of teachers to engage, mostly due to the high workload and tight classroom schedule as well as their own limited understanding of disaster risk reduction. Despite the challenges, the initiative is widely considered as a good policy intervention for formal disaster education due to its broad involvement of national governments, multilateral organizations and the civil society, ensuring comprehensive access to the target groups across an entire region and increasing sustainability of actions taken due to their alignment to other relevant national policies.

Aside of formal education, non-formal education interventions can serve as an important means in the field of disaster risk reduction. In this context, technology plays an increasingly important role. Humanitarian workers and people affected by disasters alike can be efficiently trained on disaster preparedness and management through *e-learning methodologies* such as webinar-based coaching. Even more important, however, are tools and information which can be accessed on-the-go, for instance in form of mobile applications. These can be highly effective as they are well-aligned with the very mobile nature of disasters and rescue operations. One best practice example of the use of online technologies for relief work is the *Disaster Ready Initiative*, which makes cutting-edge professional development resources available to relief and development workers and volunteers (Cornerstone on Demand Foundation, 2019).

A further good example of non-formal learning measures in Southeast Asia has been piloted by UNESCO. Following the 2011 floods in Thailand, its Bangkok office launched in 2014 the mobile game application "Sai Fah: The Flood Fighter," which teaches players flood safety lessons as they are confronted with flood hazards (UNESCO, 2019b). Available in several languages, including Bahasa Indonesian, English, French and Thai languages, it had been downloaded over 100,000 times in 140 countries by late 2015. Due to high public demand, UNESCO Bangkok subsequently developed another promising mobile phone application, in response to tsunami and earthquake risks in the region. In collaboration with the Global Disaster Preparedness Centre (GDPC) and USAID and support from the Indonesian Red Cross as well as the ASEAN Coordinating Centre for Humanitarian Assistance on Disaster Management (AHA Center), "Tanah: The Tsunami and Earthquake Fighter" was developed in 2015. The game allows to navigate interactive challenges by following the fictional girl Tanah who learns how to prepare and protect herself from tsunamis and earthquakes. According to the producers, the app allows users to learn key concepts of disaster risk reduction and enhance their hazard awareness.

As the applications developed by UNESCO have reached big numbers of mobile phone users and considering that the consumption of online data continues to be on the rise among young people worldwide, such an intervention can be considered as effective for disaster resilience education. The authors furthermore assess its acceptance as high, particularly among children and adolescents, who may feel more appealed by learning through games than formal classroom teaching.

Besides formal and non-formal disaster education, informal interventions play an important role in strengthening people's preparedness and resilience. Informal education can be a rapid entry point for disaster risk reduction education as it can be effectively implemented in engaging ways to introduce important knowledge, skills and competencies for students of all ages. Interventions can include creative educational materials such as games, documentary and short videos, cultural and performing arts such as songs, puppetry and improvization theater as well as projects bringing students in contact with local communities and governments. These practical approaches can help fostering the development of analytic and problem-solving skills among the learners.

In terms of informal education interventions, a large initiative using a peer-to-peer learning approach was implemented from 2011 to 2016 widely across South-East Asia by Plan International together with Australian Aid. The so-called "Child Centered Climate Change Adaptation (4CA)" program built the capacities of children about climate change related risks through the development and use of child-friendly education materials, such as children's books, posters, cartoons, animation films as well as games. In a particularly innovative example in Indonesia, children were trained in photography skills to identify, map and record local knowledge useful for adaptation to climate change. In a second step, children were empowered to raise awareness within their communities on disaster risk reduction and climate change adaptation. An organization-internal lessons learnt assessment revealed that local communities in the most marginalized areas, including children, are receptive to increasing their knowledge about the causes and consequences of disasters. For the learning to be effective, it is essential to build on the people's understanding and link scientific information to local wisdom (Plan International & Australian Aid, 2019).

This program can be seen as a promising example due to the low resources needed for the dissemination of knowledge, as it is members of the communities who share their learnings with others. Furthermore, it is an intervention which allows reaching out to people regardless of their educational level, including those who are not enrolled in any form of institutionalized education. Another convincing aspect of "Child Centered Climate Change Adaptation" is its strong recognition of and embedment into local realities, which can contribute to stronger acceptance of the lessons taught.

IMPLICATIONS AND POLICY LESSONS FOR DISASTER RISK REDUCTION

The examples presented above, and particularly the Association of Southeast Asian Nations (ASEAN) regional "Safe School Initiative" highlight the importance of strong governmental involvement in disaster education. Indeed, the commitment of national authorities is a key to ensure widespread access to society and especially to populations at risk. This is especially relevant in the context of formal education, considering that children from low-income households – which are more vulnerable to disasters – are likely to attend public schools. Only if the respective ministries of education broadly support the mainstreaming of disaster resilience education into national curricula, can disaster vulnerability be holistically addressed. It is therefore crucial that policy makers at all levels, including national, regional and local, take leadership in disaster education, and support capacity building for disaster preparedness. Another key lesson learnt from the initiatives described in this chapter is that there is no "one size fits all" solution to adequate disaster preparedness. As much as formal interventions are needed, their benefits such as comprehensive geographic coverage of entire regions and contact to high numbers of students are only complementary to the distinct advantages of non-formal and informal educational measures. Non-formal disaster education is particularly valuable for capacity building with groups with a low motivation or habituation for learning. Its often playful features can attract those who may not pay attention to or actively engage in structured learning, as well as people with limited literacy. Furthermore, it is useful for the immediate dissemination of life-saving information, such as in the event of a rapidly approaching disaster.

Although informal education is similarly appropriate in context of weak school performance and low reading and writing skills due to its use of creative elements, it is further a powerful tool for its strong focus on local knowledge. Taking into account that informal disaster preparedness often involves a peer-to-peer component as well as community-based learning, it represents a strong example for a bottom-up, grassroots-driven approach with high ownership of the learners. This, in turn, can strengthen the social acceptance of the educational interventions among the beneficiaries. Both non-formal and informal educational measures can moreover achieve better impacts in remote areas which are cut off from regular governmental interventions. Policy makers are thus recommended to design and support interventions which incorporate aspects of all ways of learning. Inclusiveness is another crucial factor for effective and efficient disaster education. The Malaysian case study of Plan International's "Child Centered Climate Change Adaptation" project shows that linguistic diversity in South-East Asia has to be addressed appropriately to guarantee that "no one is left behind" in disaster preparedness efforts. Not translating educational material into minority languages can be perceived as discriminatory and contribute to a widening gap of those who can take life-saving measures in emergency situations thanks to previous training and access to information and those who lack the necessary knowledge. Any disaster education strategy should hence build on equality and inclusiveness as cornerstones taking the local context and all stakeholders in the communities into account.

CONCLUSION

This chapter summarizes the growing theoretical and empirical literature on the role of education in disaster risk reduction. Education, which can take place in different environments and which can be more or less formalized, can influence disaster vulnerability in direct and indirect ways. Directly, through education and learning, individuals acquire knowledge, abilities, skills and perceptions that allow them to effectively prepare for and cope with the consequences of disasters. Indirectly, education gives individuals and households access to material, informational and social resources, which can help reducing their disaster vulnerability.

The connection between different forms of education and disaster vulnerability has been shown in various empirical studies both for the case of Southeast Asia and beyond. Better education levels were found to be positively related with various vulnerability outcomes, such as preparedness levels, reactions to early warnings, evacuations and relocation decisions, the adaptation to changing environmental conditions and the ability to cope with the consequences of disasters in the aftermath.

National risk reduction strategies have started to embrace insights from the education and learning literature to reduce the vulnerability of populations at risk. In many countries, disaster drills and courses have been integrated into national school curricula. Despite these advances, various challenges remain. Often, access to education and educational interventions is not inclusive, but strongly shaped by social, economic, political and other factors. This results in populations benefiting differently from the positive effects of education with some groups having better access to information, knowledge and other resources.

More inclusive education and information dissemination strategies, possibly combined with the usage of modern information and communication technologies, can help addressing educational inequalities. In this regard, further research and data are needed in order to better understand how the resilience of highly marginalized communities can be most effectively increased. This includes highly vulnerable populations that have otherwise limited access to public schooling and education programs, such as refugee populations. Investments in universal education and the effective and inclusive dissemination of knowledge and information can help to make substantial contributions in disaster risk reduction efforts in the Southeast Asian region.

REFERENCES

Al-Rousan, T. M., Rubenstein, L. M., & Wallace, R. B. (2014). Preparedness for natural disasters among older US Adults: A nationwide survey. *American Journal of Public Health, 104*(3), 506–511. https://doi.org/10.2105/AJPH.2013.301559

Allen, K. M. (2006). Community-based disaster preparedness and climate adaptation: Local capacity-building in the Philippines. *Disasters, 30*(1), 81–101. https://doi.org/10.1111/j.1467-9523.2006.00308.x

ASEAN (2018). Safe School Initiative. Retrieved from https://aseansafeschoolsinitiative.org/

Baker, D. P., Leon, J., Smith Greenaway, E.G., Collins, J., & Movit, M. (2011). The education effect on population health: A reassessment. *Population and Development Review, 37*(2), 307–332. https://doi.org/10.1111/j.1728-4457.2011.00412.x

Bengtsson, S., Barakat, B., & Muttarak, R. (2018). *The role of education in enabling the sustainable development agenda.* Abingon-on-Thames: Routledge.

Birkmann, J. (2008). *Assessing vulnerability before, during and after a natural disaster in Fragile Regions.* Working Paper No. 50/2008, UNU WIDER.

Blair, C., Gamson, D., Thorne, S., & Baker, D. (2005). Rising mean IQ: Cognitive demand of mathematics education for young children, population exposure to formal schooling, and the neurobiology of the prefrontal cortex. *Intelligence, 33*(1), 93–106. https://doi.org/10.1016/j.intell.2004.07.008

Card, D. (1999). The causal effect of education on earnings. In O. Ashenfelter & D. Card (Eds.), *Handbook of labor economics* (pp. 1801–1863). Amsterdam: Elsevier. https://doi.org/10.1016/S1573-4463(99)03011-4

Ceci, S. J. (1991). How much does schooling influence general intelligence and its cognitive components? A reassessment of the evidence. *Developmental Psychology*, *27*(5), 703–722. https://doi.org/10.1037/0012-1649.27.5.703

Chew, S., Heckman, J., Yi, J., Zhang, J., &bZhong, S. (2010). *Education and preferences: Experimental evidence from Chinese Adult Twins*. Conference Paper. Chicago, IL: University of Chicago.

Cornerstone on Demand Foundation. (2019). Disaster Ready Initiative. Retrieved from https://www.disasterready.org/

Cotten, S. R., & Gupta, S. S. (2004). Characteristics of online and offline health information seekers and factors that discriminate between them. *Social Science & Medicine*, *59*(9), 1795–1806. https://doi.org/10.1016/j.socscimed.2004.02.020

De Bruin, W., Parker, A. M., & Fischhoff, B. (2007). Individual differences in adult decision-making competence. *Journal of Personality and Social Psychology*, *92*(5), 938–956. https://doi.org/10.1037/0022-3514.92.5.938

Eslinger, P. J., Blair, C., Wang, J. L., Lipovsky, B., Realmuto, J., Baker, D., … Yang, O. X. (2009). Developmental shifts in FMRI activations during visuospatial relational reasoning. *Brain and Cognition*, *69*(1), 1–10. https://doi.org/10.1016/j.bandc.2008.04.010

Frankenberg, E., Sikoki, B., Sumantri, C., Suriastini, W., & Thomas, D. (2013). Education, vulnerability, and resilience after a natural disaster. *Ecology and Society*, *18*(2), 16. https://doi.org/10.5751/ES-05377-180216

Garbero, A., & Muttarak, R. (2013). Impacts of the 2010 Droughts and Floods on Community Welfare in Rural Thailand: Differential effects of village educational attainment. *Ecology and Society*, *18*(4), 27. https://doi.org/10.5751/ES-05871-180427

Grossman, M. (2006). Education and nonmarket outcomes. In E. Hanushek & F. Welch (Eds.), *Handbook of the economics of education* (pp. 577–633). Amsterdam: Elsevier. https://doi.org/10.1016/S1574-0692(06)01010-5

Guarnacci, U. (2016). Joining the dots: Social networks and community resilience in Post-Conflict, Post-Disaster Indonesia. *International Journal of Disaster Risk Reduction*, *16*, 180–191. https://doi.org/10.1016/j.ijdrr.2016.03.001

Heckman, J. J., Humphries, J.E., & Veramendi, G. (2018). Returns to education: The causal effects of education on earnings, health, and smoking. *Journal of Political Economy*, *126*(S1), 197–246. https://doi.org/10.1086/698760

Hoffmann, R., & Blecha, D. (2020). Education and disaster vulnerability in Southeast Asia: Evidence and policy implications. *Sustainability*, *12*

Hoffmann, R., & Muttarak, R. (2017). Learn from the past, prepare for the future: Impacts of education and experience on disaster preparedness in the Philippines and Thailand. *World Development*, *96*, 32–51.

Huang, J., Van den Brink, H., & Groot, W. (2009). A meta-analysis of the effect of education on social capital. *Economics of Education Review*, *28*(4), 454–464. https://doi.org/10.1016/j.econedurev.2008.03.004

Intergovernmental Panel on Climate Change (IPCC). (2014). Climate Change 2014: Synthesis Report. Contribution of Working Groups I, II and III to the Fifth Assessment Report of the Intergovernmental Panel on Climate Change. Core Writing Team, R.K. Pachauri and L.A. Meyer. Retrieved from https://doi.org/10.1017/CBO9781107415324.004

Irmansyah, I., Dharmono, S., Maramis, A., & Minas, H. (2010). Determinants of psychological morbidity in survivors of the earthquake and tsunami in Aceh and Nias. *International Journal of Mental Health Systems*, *4*, 1–10. https://doi.org/10.1186/1752-4458-4-8

Karanci, A.N., Aksit, B., & Dirik, G. (2006). Impact of a community disaster awareness training program in Turkey: Does it influence hazard-related cognitions and preparedness behaviors. *Social Behavior and Personality: An International Journal*, *33*(3), 243–258. https://doi.org/10.2224/sbp.2005.33.3.243

KC, S. (2013). Community vulnerability to floods and landslides in Nepal. *Ecology and Society*, *18*(1), 8. https://doi.org/10.5751/ES-05095-180108

Kirschenbaum, A. (2006). Families and disaster behavior: A reassessment of family preparedness. *International Journal of Mass Emergencies and Disasters*, *24*(1), 111–143.

Kohn, S., Eaton, J. L., Feroz, S., Bainbridge, A. A., Hoolachan, J., & Barnett, D. J. (2012). Personal disaster preparedness: An integrative review of the literature. *Disaster Medicine and Public Health Preparedness*, *6*(3), 217–231. https://doi.org/10.1001/dmp.2012.47

Lagmay, A. M. F., Agaton, R. P., Bahala, M. A. C., Briones, J. B. L. T., Cabacaba, K. M. C., Caro, C. V. C., … Tablazon, J. D. (2015). Devastating Storm Surges of Typhoon Haiyan. *International Journal of Disaster Risk Reduction*, *11*, 1–12. https://doi.org/10.1016/j.ijdrr.2014.10.006

Lake, R. L. D., & Huckfeldt, R. (1998). Social capital, social networks, and political participation. *Political Psychology*, *19*(3), 567–584.

Lave, T. R., & Lave, L. B. (1991). Public perception of the risks of floods: Implications for communication. *Risk Analysis*, *11*(2), 255–267. https://doi.org/10.1111/j.1539-6924.1991.tb00602.x

Lee, J. J. (2010). Review of intelligence and how to get it: Why schools and cultures count. *Intelligence*, *48*(2), 247–255. https://doi.org/10.1016/j.paid.2009.09.015

Linde, C. (2001). Narrative and social tacit knowledge. *Journal of Knowledge Management*, *5*(2), 160–171. https://doi.org/10.1108/13673270110393202

Lutz, W., Muttarak, R., & Striessnig, E. (2014). Universal education is key to enhanced climate adaptation. *Science*, *346*(6213), 1061–1062. https://doi.org/10.1126/science.1257975

Marín, A., Bodin, Ö., Gelcich, S., & Crona, B. (2015). Social capital in post-disaster recovery trajectories: Insights from a longitudinal study of tsunami-impacted small-scale fisher organizations in Chile. *Global Environmental Change*, *35*, 450–462. https://doi.org/10.1016/j.gloenvcha.2015.09.020

Marsick, V. J., & Watkins, K. E. (2002). Informal and incidental learning. *New Directions for Adult and Continuing Education*. https://doi.org/10.1002/ace.5

Muttarak, R., & Lutz, W. (2014). Is education a key to reducing vulnerability to natural disasters and hence unavoidable climate change? *Ecology and Society*, *19*(1).

Muttarak, R., Lutz, W., & Jiang, L. (2015). Introduction: What can demographers contribute to the study of vulnerability? Vienna Yearbook of Population Research, *13*, 1–13. https://doi.org/10.1553/populationyearbook2015s001

Muttarak, R., & Pothisiri, W. (2013). The role of education on disaster preparedness: Case study of 2012 Indian Ocean Earthquakes on Thailand's Andaman Coast. *Ecology and Society*, *18*(4), 51. https://doi.org/10.5751/ES-06101-180451

Neuenschwander, L. M., Abbott, A., & Mobley, A. R. (2012). Assessment of low-income adults' access to technology: Implications for nutrition education. *Journal of Nutrition Education and Behavior*, *44*(1), 60–65. https://doi.org/10.1016/j.jneb.2011.01.004

Nonaka, I., & von Krogh, G. (2009). Perspective – Tacit knowledge and knowledge conversion: Controversy and advancement in organizational knowledge creation theory. *Organization Science*, *20*(3), 481–683. https://doi.org/10.1287/orsc.1080.0412

Norris, F. H., Smith, T., & Kaniasty, K. (1999). Revisiting the experience-behavior hypothesis: The effects of Hurricane Hugo on hazard preparedness and other self-protective acts. *Basic and Applied Social Psychology*, *21*(1), 37–47 . https://doi.org/10.1207/15324839951036542

Oreopoulos, P., & Salvanes, K. G. (2011). Priceless: The nonpecuniary benefits of schooling. *Journal of Economic Perspectives*, *25*(1), 159–184. https://doi.org/10.1257/jep.25.1.159

Paton, D., & Johnston, D. (2001). Disasters and communities: Vulnerability, resilience and preparedness. *Disaster Prevention and Management: An International Journal*, *10*(4), 270–277. https://doi.org/10.1108/EUM0000000005930

Peters, E., Västfjäll, D., Slovic, P., Mertz, C. K., Mazzocco, K., & Dickert, S. (2006). Numeracy and decision making. *Psychological Science*, *17*(5), 407–413. https://doi.org/10.1111/j.1467-9280.2006.01720.x

Pichler, A., & Striessnig, E. (2013). Differential vulnerability to Hurricanes in Cuba, Haiti, and the Dominican Republic: The contribution of education. *Ecology and Society*, *18*(3). https://doi.org/10.5751/ES-05774-180331

Plan International & Australian Aid. (2019). Child Centred Climate Change Adaptation (4CA), Act to Adapt, The Next Generation Leads the Way. Retrieved from https://plan-international.org/publications/act-adapt-child-centred-climate-change-adaptation

Quartz, S. R., & Sejnowski, T. J. (1997). The neural basis of cognitive development: A constructivist manifesto. *Behavioral and Brain Sciences*, *20*, 537–596. https://doi.org/10.1017/S0140525X97001581

Reininger, B. M., Rahbar, M. H., Lee, M. J., Chen, Z., Alam, S. R., Pope, J., & Adams, B. (2013). Social capital and disaster preparedness among low income Mexican Americans in a disaster prone area. *Social Science and Medicine*, *83*, 50–60. https://doi.org/10.1016/j.socscimed.2013.01.037

Richardson, L. D., & Wolfe, M. (2001). *Principles and practice of informal education*. London: Routledge.

Rodriguez, H., Dıaz, W., Santos, J. M., & Aguirre, B. E. (2007). Communicating risk and uncertainty: Science, technology, and disasters at the Crossroads. In H. Rodriguez, E. L. Quarantelli & R. R. Dynes, *Handbook of disaster research* (pp. 476–488). New York, NY: Springer. https://doi.org/10.1007/978-0-387-32353-4_29

Russell, L. A., Goltz, J. D., & Bourque, L. B. (1995). Preparedness and hazard mitigation actions before and after two earthquakes. *Environment and Behavior*, *27*(6), 744–770. https://doi.org/10.1177/0013916595276002

Sharma, U., Patwardhan, A., & Patt, A. G. (2013). Education as a determinant of response to cyclone warnings: Evidence from coastal zones in India. *Ecology and Society*, *18*(2), 18. https://doi.org/10.5751/ES-05439-180218

Shreve, C. M., & Kelman, I. (2014). Does mitigation save? Reviewing cost-benefit analyses of disaster risk reduction. *International Journal of Disaster Risk Reduction*, *10*(A), 213–235. https://doi.org/10.1016/j.ijdrr.2014.08.004

Singh, M. (2016). *Global perspectives on recognising non-formal and informal learning: Why recognition matters. International Review of Education*. New York, NY: Springer. https://doi.org/10.1007/s11159-016-9534-0

Smith, D. L., & Notaro, S. J. (2009). Personal emergency preparedness for people with disabilities from the 2006-2007 Behavioral Risk Factor Surveillance System. *Disability and Health Journal*, *2*(2), 86–94. https://doi.org/10.1016/j.dhjo.2009.01.001

Solberg, C., Rossetto, T., & Joffe, H. (2010). The social psychology of seismic hazard adjustment: Re-evaluating the international literature. *Natural Hazards and Earth System Science*, *10*(8), 1663–1677. https://doi.org/10.5194/nhess-10-1663-2010

Thieken, A. H., Kreibich, H., Müller, M., & Merz, B. (2007). Coping with floods: Preparedness, response and recovery of flood-affected residents in Germany in 2002. *Hydrological Sciences Journal*, *52*(5), 1016–1037. https://doi.org/10.1623/hysj.52.5.1016

United Nations (UN). (2015). Transforming Our World: The 2030 Agenda for Sustainable Development. *United Nations General Assembly*. https://doi.org/10.1007/s13398-014-0173-7.2

United Nations Educational, Scientific and Cultural Organization (UNESCO). (2019a). *Functional literacy*. UNESCO Institute of Statistics. http://uis.unesco.org/node/334638

United Nations Educational, Scientific and Cultural Organization (UNESCO). (2019b). Sai Fah: The Flood Fighter. Retrieved from https://bangkok.unesco.org/content/sai-fah-flood-fighter

United Nations Office for Disaster Risk Reduction (UNISDR). (2015). Sendai Framework for Disaster Risk Reduction 2015–2030. Retrieved from https://www.unisdr.org/we/coordinate/sendai-framework

Van der Keur, P., Van Bers, C., Henriksen, H.J., Nibanupudi, H. K., Yadav, S., Wijaya, R., Subiyono, A. et al. (2016). Identification and analysis of uncertainty in disaster risk reduction and climate change adaptation in South and Southeast Asia. *International Journal of Disaster Risk Reduction*, *16*, 208–214. https://doi.org/10.1016/j.ijdrr.2016.03.002

Wamsler, C., Brink, E., & Rentala, O. (2012). Climate change, adaptation, and formal education: The role of schooling for increasing societies' adaptive capacities in El Salvador and Brazil. *Ecology and Society*, *17*(2). https://doi.org/10.5751/ES-04645-170202

Weichselgartner, J., & Pigeon, P. (2015). The role of knowledge in disaster risk reduction. *International Journal of Disaster Risk Science*, *6*(2), 107–116. https://doi.org/10.1007/s13753-015-0052-7

Wen, L. M., Rissel, C., Baur, L. A., Lee, E., & Simpson, J. M. (2011). Who is NOT likely to access the internet for health information? Findings from first-time mothers in Southwest Sydney, Australia. *International Journal of Medical Informatics*, *80*(6), 406–411. https://doi.org/10.1016/j.ijmedinf.2011.03.001

Werquin, P. (2010). *Recognising non-formal and informal learning: outcomes, policies and practices*. OECD Publishing. https://doi.org/10.1787/9789264063853-en

Witvorapong, N., Muttarak, R., & Pothisiri, W. (2015). Social participation and disaster risk reduction behaviors in Tsunami Prone Areas. *PLoS ONE*, *10*(7), 1–20. https://doi.org/10.1371/journal.pone.0130862

World Bank. (2019). *World development indicators*. Washington, DC: World Bank.

Xiao, C., & McCright, A. M. (2007). Environmental concern and sociodemographic variables: A study of statistical models. *Journal of Environmental Education*, *38*(2), 3–14. https://doi.org/10.3200/JOEE.38.1.3-14

Zagheni, E., Muttarak, R., & Striessnig, E. (2015). Differential mortality patterns from hydro-meteorological disasters: Evidence from cause-of-death data by age and sex. *Vienna Yearbook of Population Research*, *13*, 47–70. https://doi.org/10.1553/populationyearbook2015s047

CHAPTER 3

ENABLING ORGANIZATIONAL RESILIENCE AMIDST DISASTERS: EXPERIENCES OF WORK ORGANIZATIONS IN SOUTHEAST ASIA

Mendiola Teng-Calleja, Pinky Rose Sabile and Angelique Pearl Virtue Villasanta

ABSTRACT

This chapter describes the environmental risks and vulnerabilities faced by work organizations in Southeast Asia. It also presents cases that demonstrate how these organizations respond to disasters and natural hazards. To situate the case discussions, a review of existing studies of organizational resilience, particularly those that propose definitions, models, and frameworks is presented. The cases from the Philippines and Thailand illustrate how active and integrative efforts at building resilience can be institutionalized at the organizational level.

Keywords: Work organizations; organizational resilience; disasters; vulnerabilities; Southeast Asia; risks

This chapter describes the environmental risks faced by work organizations in Southeast Asia and how these organizations build resilience against disasters and natural hazards. To situate this in literature, the following section describes

Resistance, Resilience, and Recovery from Disasters: Perspectives from Southeast Asia
Community, Environment and Disaster Risk Management, Volume 21, 39–52

ISSN: 2040-7262/doi:10.1108/S2040-726220200000021003

the concept of risks and vulnerabilities. This is followed by a review of existing studies on organizational resilience, particularly those that propose definitions, models, and frameworks. This chapter ends with a presentation of cases that outline risks and vulnerabilities experienced by organizations located in this area as well as demonstrate the current state of resilience efforts in work organizations. Some of the cases likewise illustrate how active and integrative efforts at building resilience can be institutionalized at the organizational level.

RISKS AND VULNERABILITIES

The United Nations International Strategy for Disaster Reduction (UNISDR, 2004) describes risks as having two essential elements – hazard and vulnerability. Hazard is defined as

> a dangerous phenomenon, substance, human activity or condition that may cause loss of life, injury or other health impacts, property damage, loss of livelihoods and services, social and economic disruption, or environmental damage. (UNISDR, 2009, p. 17)

Vulnerability, on the other hand, pertains to the lack of ability to endure or withstand the effects of adverse environmental conditions (Ciurean, Schroter, & Glade, 2013). Assessing vulnerability of work organizations entails looking at the characteristics of society and the environments that may make them predisposed to incurring losses and damages. Gallopin (2012) expanded this definition of vulnerability to include not just the exposure to external hazards and the degree to which a system may be disturbed by the adverse situation (i.e., sensitivity to the perturbation), but also the capacity of a system, such as an organization, to respond. As shown in Fig. 1, Gallopin's (2012) model placed resilience or the capacity to recover from a hazard as a subset of vulnerability (the more encompassing concept) together with adaptive capacity (Bhamra, Dani, & Burnard, 2011).

There are a number of definitions and frameworks that describe resilience as a distinct construct. In the succeeding section, we present organizational resilience definitions and frameworks particularly those applied to work organizations.

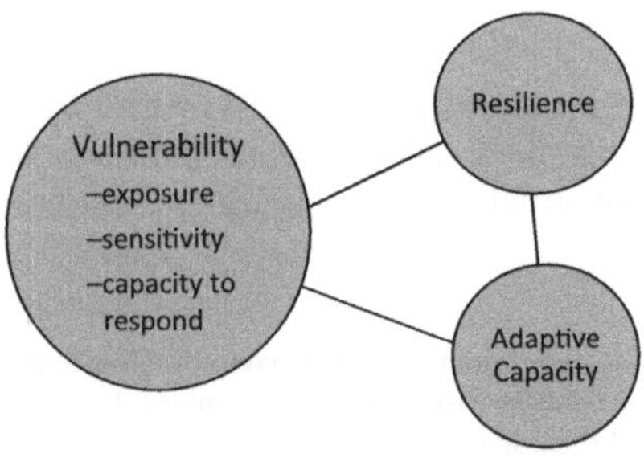

Fig. 1. Conceptualization of Vulnerability. Adapted from Gallopin (2006).

ORGANIZATIONAL RESILIENCE

Work organizations are currently faced with turbulent, disruptive, and threatening environments where occasional jolts or even unexpected occurrences are ever-present (Weick & Sutcliffe, 2007). Disasters or hazards can destabilize the operations and security of organizations (Bhamra et al., 2011). Organizational systems are thus exposed to wider environmental risks that if not addressed effectively will make the work organization's future both uncertain and unpredictable (Burnard & Bhamra, 2011).

Given the potential for disruption, it is important for work organizations to be prepared for the unexpected. Preparation may mean developing resilience elements such as adaptive organizational structures that will enable risk tolerance, and an innate ability to proactively respond to environmental uncertainty (Burnard & Bhamra, 2011).

Awareness of environmental hazards is likewise critical in developing organizational resilience. McManus (2008) cited in Lee, Vargo, and Seville (2013) highlighted the importance of awareness in proposing a definition constructed through an empirical study involving organizations in New Zealand. Specifically, she defined organizational resilience as "a function of an organization's overall situation awareness, management of keystone vulnerabilities and adaptive capacity in a complex, dynamic and interconnected environment" (p. 82). Situation awareness includes knowing and analyzing hazards as well as its consequences. Awareness intuitively precedes the succeeding two dimensions – managing keystone vulnerability and adaptive capacity. Managing vulnerabilities involves planning strategies, participating in exercises, and strengthening the capacity and capabilities of internal and external resources. Adaptive capacity means breaking silo mentalities, having a strategic vision, enhancing communications, relationships, information and knowledge sharing as well as leadership, management and structures for governance (McManus, 2008 as cited in Lee et al., 2013).

Organizational resilience is more commonly defined as the ability to recover from unexpected, stressful, and adverse situations (e.g., Dutton, Frost, Worline, Lilius, & Kanov, 2002; Gittell, Cameron, Lim, and Rivas, 2006; Rudolph & Repenning, 2002). This perspective highlights the rebound-oriented mechanisms of work organizations through different coping strategies (Lengnick-Hall & Beck, 2005). However, Beuren and dos Santos (2019) characterized this as a passive approach to organizational resilience. An approach that reflects a more active stance implies that work organizations need to develop stronger capacities and a range of new actions prior to experiencing a hazard or a disruptive event (Lengnick-Hall, Beck and Lengnick-Hall; Somers, 2009).

Lengnick-Hall and Beck (2005) and Lengnick-Hall, Beck, and Lengnick-Hall (2011) described this active approach to organizational resilience as resilience capacity that combines cognitive, behavioral, and contextual capacities at the organizational level. These authors suggest that an organization that has resilience capacity will be able to more creatively interpret ambiguous and disruptive situations (cognitive resilience), will have greater capability to come up with unconventional responses (behavioral resilience), and maximize available resources and partnerships/networks (contextual relationships). Lengnick-Hall and Beck (2005) further described the four conditions of psychological safety, deep social capital, diffused power and accountability, and broad resource network as contextual factors facilitative of effective disaster response. Psychological safety describes the extent by which people perceive their work environment as conducive to interpersonal risk-taking. Deep social capital grows through positive interactions within the work organization. Diffused power and accountability happens when self-organization is encouraged and the work organization is not managed hierarchically. Access to broad networks entails building relationships with others and forming strategic alliances.

Organizational resilience may also mean developing capabilities necessary to thrive despite challenging circumstances. Capabilities, in the context of organizational resilience, may pertain to endowments that shape the ability of work organizations to make sense of and respond to unfamiliar challenges in a positive way (Sutcliffe & Vogus, 2003). These endowments can be in the form of cognitive, behavioral, emotional, or relational capabilities. Cognitive capability pertains to having a shared purpose and strong values. It also refers to the knowledge and expertise possessed by members of the work organization that they can utilize in the face of adversity (Lengnick-Hall et al., 2011). Behavioral capabilities include the range of alternative behaviors enabled by organizational structure and processes that facilitate information sharing as well as cooperation and coordination (Buchwald & Hobfoll, 2004). Emotional-regulation means mental hardiness and the ability to regulate one's self to cope with adversity and remain productive while relational capability are the social connections that facilitate exchange and enable access to resources (Buchwald & Hobfoll, 2004).

A similarly active approach that incorporates an integrative and longitudinal perspective (Nucifora, Hall, & Everly, 2011) of organizational resilience is the John Hopkins Resistance–Resilience–Recovery Model. The model does not only focus on what happens immediately after the occurrence of a disaster/hazard (resilience), but also to what occurs before (resistance) and during the period of going back to normalcy (recovery).

The model described resistance as the capacity of the organization and individuals within it to withstand distress caused by disasters (Nucifora, Langlieb, Siegal, Everly, & Kaminsky, 2007). This includes proactive interventions done prior to experiencing the disaster that build psychological immunity. Resilience enhancing interventions address conditions or challenges immediately after the disaster. This pertains to initiatives that promote the ability to rebound from the psychological distress and various forms of destruction resulting from the calamity (Nucifora et al., 2011). Recovery pertains to the provision of treatment and rehabilitation that aim to enhance organizational capacity to adapt and function despite experiencing massive disruptions and maladjustments (Nucifora et al., 2011).

Work organizations sustain losses when disasters strike. However, there seems to be a dearth of studies on resilience in these kinds of organizations. Much of what is known about organizational resilience is based on initiatives in Japan and New Zealand, two disaster-prone high-income countries (e.g., Crowley & Elliott, 2012). Even though the Asia-Pacific region (that includes Japan, New Zealand, and Southeast Asia) as a whole is generally regarded as part of the most disaster-prone area in the world (ESCAP, 2016), not all countries in this region have the same economic resources as high-income countries. Scarcity of resources can limit strategies of promoting resilience among work organizations (ESCAP, 2013). Given this context, this chapter asks how do Southeast Asian organizations build resilience?

RISKS, VULNERABILITIES, AND ORGANIZATIONAL RESILIENCE

Examples from Southeast Asian Work Organizations

In the face of natural hazards and disasters, work organizations become vulnerable to damages and losses that may threaten their very existence. For instance, 10% of work organizations affected by the 2010 Pakistani floods inevitably closed down. The remaining 90% that reopened after the flooding depended on external sources of support to be able to recover from their losses (Asgary et al., 2012, as cited in Chatterjee, Ismail, & Shaw, 2016). In 2011, 656 work organizations in Japan were unable to recover from losses and disruption caused by the East Japan earthquake and tsunami (ADRC, 2012, as cited in Chatterjee et al., 2016). Compounding financial losses from disasters is the additional pressure for work organizations to continue providing livelihood or to extend financial assistance to their employees and dependents.

The magnitude of exposure and sensitivity of work organizations in Southeast Asia to external hazards as well as their responses to these vulnerabilities are further described through the following cases: (1) the 2004 Indian Ocean tsunami, (2) Vietnam's typhoons, (3) 2011 Thailand floods, and (4) disasters and emergencies in the Philippines. Some of the documented responses regarding preparations for natural hazards were based on the survey conducted by iPrepare Business (2017a, 2017b, 2017c) on the disaster resilience of small and medium enterprises (SMEs) in Asia.

THE 2004 INDIAN OCEAN TSUNAMI

Following a 9.1 magnitude earthquake beneath the Indian Ocean, a massive tsunami struck and claimed more than 230,000 lives across 10 different countries (Taylor, 2014). It damaged coastal areas and villages of Thailand, Sri Lanka, India, Malaysia, Maldives, and Indonesia. Indonesia was recorded to have sustained the strongest economic impact, losing an estimated 4.45 billion USD, 60% of which (or 2.67 billion USD) were due to direct damages, and 40% (or 1.78 billion USD) were losses due to disrupted income flow (Athukorala & Resosudarmo, 2005). Assets, infrastructures, and natural resources were destroyed, leading to disruption in production, higher operational costs, and costs associated with reconstruction, relocation, and/or recovery (Asian Disaster Preparedness Center, n.d.). Following this disaster, initiatives to improve tsunami early warning systems were put in place. Trainings to increase public awareness were also conducted. Aside from these, infrastructures along coastlines were designed and built to withstand tsunamis.

Despite the grave impact of the 2004 tsunami on Indonesia, Indonesian work organizations reported lower awareness of risks brought about by natural hazards compared to other organizations in Southeast Asia (iPrepare Business, 2017a). Basing it on McManus' framework (2008, as cited in Lee et al., 2013) situational awareness is a critical component of organizational resiliency. This shapes how the organization manages vulnerabilities as well as develops adaptive capacity. The relationship among these three dimensions of organizational resilience may perhaps explain why only 14% of Indonesian SMEs reported having a business continuity plan (BCP) or a documented procedure that details how to get the work organization running again after experiencing a hazard or crisis (Queensland Government, 2017). Preparing a strategy is akin to having a BCP and is a component of managing keystone vulnerabilities. Having a strategy may then shape other efforts to strengthen the adaptive capacity of the organization.

Lack of awareness of exposure and sensitivity to vulnerabilities were likewise reported by work organizations in Vietnam that responded to the 2017 iPrepare business survey. Looking at it from both Gallopin (2012) and McManus' (2008, as cited in Lee et al., 2013) frameworks, the lack of awareness appeared to have weakened the capacity of work organizations to prepare for natural hazards.

VIETNAM

Disasters, such as typhoons, storms, and flooding are yearly threats to Vietnam. Vietnam experiences an average of 10 storms per year, claiming the lives of more than 400 people. Because of its long coastline and the presence of several river deltas (particularly the Mekong Delta and the Red River Delta), Vietnam is also vulnerable to sea-level rises. Flooding in urban areas is likewise frequent. However, like in Indonesia, work organizations in Vietnam rarely used BCPs and reported relatively low levels of awareness on risks brought about by disasters (iPrepare Business, 2017c). Business disruption was most often caused by inability of employees to report to work, interferences in the delivery of products to consumers, and damages to facilities and materials (iPrepare Business, 2017c). It must be noted that organizations that were surveyed were relatively young, having been established after 2009, and may have limited experience of dealing directly with disasters. Previous studies found that prior experience of disaster is one of the enablers of disaster preparedness (Hoffmann & Muttarak, 2017).

THE 2011 THAILAND FLOODS

After five months of monsoon rains, Thailand experienced severe flooding in its cities in 2011. The rains have claimed more than 800 lives, displaced millions, and destroyed houses and infrastructures (European Space Agency, 2011; Impact Forecasting, AON Benfield, 2012). Prolonged rains and flooding have also forced about 1,000 factories to close, with total damage to industry estimated to reach from 3.3 billion USD to as high as 45.7 billion USD (British Broadcasting Corporation (BBC), 2011; Impact Forecasting, AON Benfield, 2012). In general, work organizations in Thailand were found to be greatly affected by disasters as these lead to lengthy work disruptions. The 2011 floods, in particular, were termed the second costliest Asian disaster in the last decade. Seven major industrial states were submerged in floodwaters as high as 10 feet. As of 2017, only 15% of small enterprises and 21.5% of medium enterprises had prepared BCPs (iPrepare Business, 2017a) indicating a relatively low awareness of vulnerabilities. However, this was a great leap from a 2012 survey that found all SME's in Thailand not having a BCP (iPrepare Business, 2017a). The report noted that one reason given by organizations for not having a BCP was they are unlikely to encounter a disaster any time soon given that they had just experienced one.

These observations among SMEs were very different from the experience of Western Digital (WD) company located in the Bang Pa-in Industrial Estate in Thailand. Despite considerable damages in its production equipment, WD was able to unexpectedly resume operations 46 days after the massive flooding on October 15, 2011 (Haraguchi & Lall, 2014). This was less than half the time required by a competitor organization that experienced the same level of damages in the 2011 flooding (Haraguchi & Lall, 2014). WD's Vice President for Magnetic Operations, Sampan Silapanad claimed that "The recovery has proven that WD employees have exceptional passion and perseverance" and that they "did everything to make the plant equal or better than the pre-flood condition" (Sabayjai, 2012, p. 2, as cited in Bunya & Tang, 2013).

How WD was able to rise as a better work organization after the 2011 flooding reflects an active approach to building organizational resilience (Lengnick-Hall & Beck, 2005; Lengnick-Hall et al., 2011) and demonstrates the components of the Johns Hopkins Resistance–Resilience–Recovery model.

Prior to the flooding, WD monitored and assessed risks factors (e.g., rainfall and floodwater levels) as well as took proactive measures to protect the plant from floodwaters by preparing generators, sandbags, water pumps, food, and first-aid kits (Bunya & Tang, 2013). They likewise developed mitigation, contingency, and communication plans. During the disaster, continuous monitoring of employees as well as the flood situation was done by the BCP team and management team. These teams likewise facilitated the execution of plans to reduce the time needed to recover submerged equipment and minimize disruption in their supply chain as well as ensure the well-being and safety of employees (Bunya & Tang, 2013). Aside from the resistance and resilience efforts, WD likewise engaged in recovery initiatives by restoring production equipment (80% of which were made functional again) and instituting a decontamination process of its facility. Internal communications among its work units and external stakeholders was likewise focused on during recovery (Bunya & Tang, 2013). The resistance, resilience, and recovery initiatives of WD enabled it to efficiently resume business operations after experiencing disaster. Considerable sales volume in the succeeding year also showed greater confidence of customers in the organization and its products (Bunya & Tang, 2013).

The last case that will be presented describes risks and vulnerabilities as well as efforts at building organizational resilience among work organizations in the Philippines. The country was identified by the United Nations Office for Disaster Risk Reduction (2015) as one of the most disaster-prone countries in the world. It also ranked third in the 2016 World Risk Index since it frequently "experiences almost all forms of natural and human-induced calamities, such as typhoons, earthquakes, floods, volcanic eruptions, landslides, fires and armed conflict" (https://philippines.unfpa.org). Like Thailand, the case provides more detailed examples of how work organizations in the Philippines endeavor to achieve organizational resilience.

THE PHILIPPINES

Based on the 2017 iPrepare Business survey conducted among SMEs in Asia, work organizations located in the Philippines have relatively higher levels of awareness when it comes to disaster risks compared to their counterparts in Southeast Asia. This may be attributed to the frequent occurrences of disasters in the country such as earthquakes, typhoons, cyclones, and flooding (Wingard & Brändlin, 2013). Work organizations are also affected by disasters in varying degrees. Impact of disaster ranges from financial losses to work disruptions that may sometimes be as short as less than a day, to longer periods of disruption spanning to more than three months (iPrepare Business, 2017b). While awareness regarding disaster risk may be high, only a few of the SMEs (6%) have an existing formal BCPs. Reasons for not preparing a BCP are mainly due to lack of awareness about BCPs and lack of information or expertise to prepare one (iPrepare Business, 2017b). This shows that apart from disaster awareness, endowments such as cognitive capability or having knowledge and expertise that can be utilized in the face of adversity (Lengnick-Hall et al., 2011) is necessary in building organizational resilience.

A multiple-case study conducted by Teng-Calleja, Hechanova, Sabile, and Villasanta (2020) in the Philippines demonstrate practices that reflect an active approach taken by six organizations toward building organizational resilience. Similar to the experience of WD in Thailand, the efforts of work organizations that participated in the research reflected the longitudinal perspective of the Johns Hopkins Resistance–Resilience–Recovery model.

All six organizations have experienced disasters such as massive flooding due to tropical cyclones (e.g., Typhoon Haiyan in 2013) and heavy monsoon rains (*Habagat* in Tagalog) that occur almost annually in the country. Some have also experienced earthquakes (e.g., in Bohol and adjacent areas last 2018) and human-induced perturbations such as fires, oil spills, and even bomb threats. Members of these organizations shared that damage to property and financial losses was the most immediate impact of disasters. Experiencing hazards also affected the well-being and productivity of employees.

All six organizations put in place systems, structures, and/or resources that helped their organization not only to withstand the negative effects of disasters but also to more efficiently withstand adversity. As part of efforts to develop resistance, the work organizations built information and communication systems to monitor employees' situation as soon as disasters happen. They also created active employee databases that will allow them to reach out to all employees. Two of the companies have prepared communication contingency equipment such as satellite phones and radio systems. The work organizations also instituted protocols (e.g., work suspension, ways to determine the intensity of the emergency situation and the level of response needed, as well as protocols for addressing reputational risks) to enable immediate and appropriate response. They likewise formed and trained emergency response teams.

All six organizations have either a BCP or an articulated strategy that they roll-out when disaster happens. They have also allocated resources such as calamity fund for emergencies and building maintenance and repair, invested in rescue vehicles, big trucks, and boats as well as identified, marked, and communicated assembly areas as well as evacuation sites to their employees and other stakeholders. To further enhance their resistance-building initiatives and their capacity to manage the impact of disasters, some of the organizations have forged partnerships with other volunteer organizations as well as national and local government organizations for safety exercises, drills, first-aid training, and mutual support, especially in rescue efforts upon the occurrence of natural hazards. It was also interesting that one of the schools partnered with a drugstore and a supermarket in front of their campus so they may have access to medicines, food, and other needs when disaster strikes.

In addition to these efforts, some organizations provided their employees with go-bags (a portable bag that contains things that you need to survive in an emergency situation such as water, flashlight, first-aid kit, whistle, and nutrition bars) while others shared information on what the go-bags need to contain. There are also those that distributed hard hats, smoke masks, and other fire protection equipment to employees. The organizations likewise invested in developing disaster-relevant knowledge and skills among their employees. All of the organizations have practice drills or dry-runs of systems and protocols for different types of emergencies (e.g., earthquake, fire, bomb threats) and some even evaluates the drills to determine areas for improvement. Aside from formal training, some of the organizations also provide avenues for story telling (i.e., sharing of stories on how to cope with disasters) as well as ideation exercises (or idea generation activities done with groups of employees) to muster creative ways of dealing with disasters. These resistance-building initiatives of the six organizations reflect the cognitive, behavioral, and contextual resilience enhancing capacities described by Lengnick-Hall and Beck (2005) and Lengnick-Hall et al. (2011).

Immediately after the occurrence of disasters, the organizations expect to activate the systems, protocols, and structures especially for decision-making and provision of response and rescue operations. They likewise conduct an impact assessment to immediately account for their employees as well as the employees' families and monitor damage to their facilities. As part of their resilience-building the work organization also provide psychological safeguards and support (e.g., knowledge that the organization will rescue employees trapped in flooded areas) as well as financial and material resources (e.g., water for drinking, food, medicine, candles, and blankets) to employees. The provision of resources sometimes extends to the employees' families, relatives, and neighbors.

There seems to be fewer recovery programs compared to resistance and resilience-building initiatives in the six organizations. Aside from having BCPs that enable the immediate rebuilding of facilities and restoration of operations and service, some organizations also make sure that they build strong partnerships with people and establishments in the community where they operate. Some organizations also continuously provide emotional support to their employees and relaxed the rules regarding attendance and deliverables especially for those needing further psychological support. Affected employees are likewise provided financial support and loans with flexible payment schemes.

A more detailed summary of the organizational resilience initiatives and programs of the six work organizations in the Philippines following the Johns Hopkins Resistance–Resilience–Recovery model is presented in Appendix. The summary can be used by work organizations in Southeast Asia in assessing their resilience-building initiatives.

CONCLUSIONS

This chapter highlights the need for most work organizations in Southeast Asia to proactively identify and implement more targeted resilience-building interventions. This may include efforts to craft plans that will ensure organizational sustainability and health (e.g., through BCPs). These initiatives need to consider varied aspects such as level of exposure and sensitivity to vulnerabilities that may shape appropriate response. Practices and programs shared in this chapter may guide organizational resilience-building efforts as these may be more contextually nuanced than those practices in high-income countries. However, there is still a need to document/include experiences of organizations in other Southeast Asian countries and consolidate more knowledge on organizational resilience-building practices in the region. The challenge is to enable research efforts at the regional level which can inform practice in the entire region especially since disasters sometimes affect multiple countries (e.g., 2004 Indian Ocean tsunami). Governments in the Association of Southeast Asian Nations (ASEAN) may also explore standard regulatory practices that may guide (if not push) resilience-building efforts in work organizations.

APPENDIX

Summary of the Organizational Resilience Initiatives and Programs of the Six Work Organizations in the Philippines Following the Johns Hopkins Resistance–Resilience–Recovery Model

	Resistance
a. Information and communication systems	• Communication system for warning employees • System for communication with all employees • Back-up communication systems (e.g., satellite phones) • System to account for employees • Database of employees who are living in disaster-prone areas • Mechanisms for monitoring threats (e.g., weather-related threats; security-related threats)
b. Protocols	• Protocols regarding work suspension • Protocols on responding to reputational risk (e.g., activation of Public Relations team) • Protocols regarding deployment of response teams • Protocols regarding activation of rescue operations • Protocols regarding activation of emergency states or alert levels
c. Formation and training of response teams	• Emergency response team • Training of emergency response team (e.g., first-aid training, firefighting exercises, swimming)
d. BCP	• A documented BCP • A group in-charge of the BCP • A unit responsible for quick decision-making during emergencies
e. Allocating resources	• A calamity fund • Rescue vehicles • Prepared infrastructures (e.g., retrofitting buildings, repairing, stabilization efforts) • Transportation services for employees in case of emergencies • Identified assembly/evacuation area for employees
f. Building a network of experts and partners	• Partnerships with outside institutions or experts on emergency situations (e.g., external response teams) • Partnership with national agencies • Partnerships with local government units • Partnership with resource providers (e.g., drugstore, supermarket)
g. Building disaster-relevant knowledge and skills	• Practice drills • Access to disaster information • Awareness of the fundamental disaster processes at work • Knowledge of where the emergency exit is • Active participation during drills and simulation • Opportunity to share best practices during emergency planning • Learning sessions on disaster preparedness • Awareness on the available disaster-trainings and development at work
h. Providing resources	• Access to disaster preparedness kits • Access to safety equipment at work • Knowledge on where the emergency equipment are
	Resilience
a. Activation of systems, protocols, and structures	• Adherence to protocols on BCP • Monitoring mechanisms if protocols are followed • Activation of unit that is responsible for quick decision-making during emergencies • Deployment of emergency response teams • Rescue of employees in need

(Continued)

(Continued)

b. Impact assessment	• Checking facilities postdisaster • Assessment of postdisaster needs of employees • Assessment of the impact and damages • Checking the safety of employees • Accounting for its employees
c. Psychological safeguards and support	• Allowing employees in affected areas to stay at home with their families during disaster • Programs and benefits provided to victims of disaster • Responders and rescue efforts when disaster strikes • Provision of psychological support during disaster (i.e., psychological first aid [PFA], debriefing)
d. Material and financial resources	• Donations from the company (food, clothing) when disaster strikes • Provision of food, water, clothing, and/or shelter for employees their families, and neighbors affected by the disaster • Financials assistance to employees • Financials assistance to families of employees
	Recovery
a. Postdisaster Psychological Support	• Leaves or work from home arrangement postdisaster • Access to counseling sessions at work • Access to emotional support from colleagues at work
b. Evaluation of protocols	• Assessment of flaws in the protocols • Assessment of flaws in the execution of protocols
c. Restoration/ Rebuilding of Facilities	• Rebuild/repair of damaged facilities • Clean up of affected facilities • Strategies to build stronger facilities • Mitigation measures against future damages to facilities

REFERENCES

AON Benfield. (2012). September 2012 global catastrophe recap. Retrieved from http://thoughtleadership.aonbenfield.com/Documents/20121004_if_global_cat_recap_september.pdf

Asian Disaster Preparedness Center. (n.d.). Social and economic impact of December 2004 tsunami. Retrieved from http://cmsdata.iucn.org/downloads/social_and_economic_impact_of_december_2004_tsunami_apdc.pdf

Athukorala, P., & Resosudarmo, B. (2005). The Indian Ocean tsunami: Economic impact, disaster management and lessons. Departmental Working Papers 2005-05, The Australian National University, Arndt-Corden Department of Economics.

British Broadcasting Corporation (BBC). (2011, October 20). Thailand floods: Bangkok 'impossible to protect'. BBC News. Retrieved from https://www.bbc.com/news/world-asia-pacific-15381227

Bhamra, R., Dani, S., & Burnard, K. (2011). Resilience: The concept, a literature review and future directions. *International Journal of Production Research*, *49*(18), 5375–5393.

Beuren, I. M., & dos Santos, V. (2019). Enabling and coercive management control systems and organizational resilience. *Revista Contabilidade & Finanças*, *30*(81), 307–323. doi:10.1590/1808-057x201908210

Buchwald, P., & Hobfoll. S. (2004). Burnout in the conservation of resources theory. *Psychologie in Erziehung und Unterricht (PEU)*, *51*(4), 247–257. doi:10.1016/B978-012373947-6.00093-3

Bunya, & Tang (2013). Teaching Case UCLA Anderson School of Management. Retrieved from http://www.anderson.ucla.edu/documents/areas/fac/dotm/bio/V4EDITResilientResponseRecoveryWD.pdf

Burnard, K., & Bhamra, R. (2011). Organisational resilience: Development of a conceptual framework for organisational responses. *International Journal of Production Research*, *49*(18), 5581–5599.

Chatterjee, R., Ismail, N., & Shaw, R. (2016). Identifying priorities of Asian small- and medium-scale enterprises for building disaster resilience. In R. Shaw, A. Rahman, A. Surjan, & G. A. Parvin (Eds.), *Urban disasters and resilience in Asia* (pp. 179–194). London: Elsevier.

Ciurean, R. L., Schroter, D., & Glade T. (2013). Conceptual frameworks of vulnerability assessments for natural disasters reduction. In J. Tiefenbacher (Ed), *Approaches to disaster management: Examining the implications of hazards, emergencies and disasters*. InTech. http://dx.doi.org/10.5772/55538

Crowley, K., & Elliott, J. R. (2012). Earthquake disasters and resilience in the global North: Lessons from New Zealand and Japan. *The Geographical Journal, 178*(3), 208–215.

Dutton, J. E., Frost, P. J., Worline, M. C., Lilius, J. M., & Kanov, J. M. (2002). Leading in times of trauma. *Harvard Business Review, 80*(1), 54–61.

European Space Agency. (2011). Flood – Thailand, October 2011. Retrieved from https://earth.esa.int/web/earth-watching/natural-disasters/floods/content/-/article/flood-thailand-october-2011

ESCAP. (2013). Regional issues in disaster risk reduction, including those related to climate change adaptation, and policies related to mainstreaming disaster risk reduction into socio-economic development planning. Information Communication Technology and Disaster Risk Reduction Section. Retrieved from https://www.unescap.org/sites/default/files/Regional%20issues%20in%20DRR%2C%20including%20those%20related%20to%20CCA.pdf

ESCAP. (2016). Disaster in Asia and the Pacific: 2015 year in review. Retrieved from https://www.unescap.org/sites/default/files/2015_Year%20in%20Review_final_PDF_1.pdf

Gallopin, G. C. (2006). Linkages between vulnerability, resilience, and adaptive capacity. *Global Environmental Change, 16*(3), 293–303.

Gittell, J. H., Cameron, K., Lim, S., & Rivas, V. (2006). Relationships, layoffs, and organizational resilience: Airline industry responses to September 11. *The Journal of Applied Behavioral Science, 42*(3), 300–329.

Haraguchi, M., & Lall, U. (2014). Flood risks and impacts: A case study of Thailand's floods in 2011 and research questions for supply chain decision making. *Journal of Disaster Risk Reduction*, 1–16. doi:10.1016/j.ijdrr.2014.09.005i

Hoffmann, R., & Muttarak, R., (2017). Learn from the past, prepare for the future: Impacts of education and experience on disaster preparedness in the Philippines and Thailand. *World Development, 96*, 32–51. doi:10.1016/j.worlddev.2017.02.016

Impact Forecasting, AON Benfield. (2012). 2011 Thailand floods event recap report: Impact forecasting – March 2012. Retrieved from http://thoughtleadership.aonbenfield.com/Documents/20120314_impact_forecasting_thailand_flood_event_recap.pdf

iPrepare Business. (2017a). Strengthening disaster and climate resilience of small & medium enterprises in Asia. Retrieved from https://app.adpc.net/publications/regional-synthesis-report-strengthening-disaster-resilience-small-and-medium

iPrepare Business. (2017b). Strengthening disaster and climate resilience of small & medium enterprises in Asia: Philippines. Retrieved from https://app.adpc.net/sites/default/files/public/publications/attachments/2%20IP_SR_Philippines.pdf

iPrepare Business. (2017c). Strengthening disaster and climate resilience of small & medium enterprises in Asia: Viet Nam. Retrieved from https://www.researchgate.net/publication/320554389_enabling_environment_opportunities_viet_nam_strengthening_disaster_and_climate_resilience_of_small_medium_enterprises_in_asia

Lee, A.V., Vargo, J., & Seville, E. (2013). Developing a tool to measure and compare organizations' resilience. *Natural Hazards Review, 14*(1), 29–41.

Lengnick-Hall, C. A., & Beck, T. E. (2005). Adaptive fit versus robust transformation: How organizations respond to environmental change. *Journal of Management, 31*(5), 738–757.

Lengnick-Hall, C. A., Beck, T. E., & Lengnick-Hall, M. L. (2011). Developing a capacity for organizational resilience through strategic human resource management. *Human Resource Management Review, 21*(3), 243–255.

Nucifora, F., Hall, R., & Everly, G. (2011). Reexamining the role of the traumatic stressor and the trajectory of posttraumatic distress in the wake of disaster. *Disaster Medicine and Public Health Preparedness, 5*(2). doi:10.1001/dmp.2011.51

Nucifora, F., Langlieb, A., Siegal, E., Everly, G., & Kaminsky, M. (2007). Building resistance, resilience, and recovery in the wake of school and workplace violence. *Disaster Medicine and Public Health Preparedness, 1*(1). doi:10.1097/DMP.0b013e31814b98ae

Queensland Government. (2017). What's in a business continuity plan? Retrieved from https://www.business.qld.gov.au/running-business/protecting-business/risk-management/continuity-planning/plan

Rudolph, J. W., & Repenning, N. (2002). Disaster dynamics: Understanding the role of quantity in organizational collapse. *Administrative Science Quarterly*, *47*(1), 1–30. doi:10.2307/3094889

Somers, S. (2009). Measuring resilience potential: An adaptive strategy for organizational crisis planning. *Journal of Contingencies and Crisis Management*, *17*(1), 12–23.

Sutcliffe, K. M., & T. J. Vogus (2003). Organizing for resilience. In K. S. Cameron, J. E. Dutton, & R. E. Quinn (Eds), *Positive organizational scholarship: Foundations of a new discipline* (pp. 94–110). San Francisco, CA: Berrett-Koehler.

Taylor, A. (2014, December 26). Ten years since the 2004 Indian Ocean tsunami. *The Atlantic*. Retrieved from https://www.theatlantic.com/photo/2014/12/ten-years-since-the-2004-indian-ocean-tsunami/100878/

Teng-Calleja, M., Hechanova, M. R. M., Sabile, P. R., & Villasanta, A. P. V. P. (2020). *Building organization and employee resilience in disaster contexts. International Journal of Workplace Health Management*. Advance online publication. doi 10.1108/IJWHM-09-2019-0122.

United Nations International Strategy for Disaster Reduction (UNISDR). (2004). Living with risk: A global review of disaster reduction initiatives. Retrieved from https://www.undrr.org/publication/living-risk-global-review-disaster-reduction-initiatives

United Nations International Strategy for Disaster Reduction (UNISDR). (2009). Terminology on disaster risk reduction. Retrieved from http://www.unisdr.org/files/7817_UNISDRTerminologyEnglish

United Nations Office for Disaster Risk Reduction. (2015). The human cost of weather-related disasters (1995-2015). Retrieved from https://www.undrr.org/publication/human-cost-weather-related-disasters-1995-2015

Wingard, J., & Brändlin, A. (2013, November 10). Philippines: A country prone to natural disasters. *Deutsche Welle*. Retrieved from https://www.dw.com/en/philippines-a-country-prone-to-natural-disasters/a-17217404

Weick, K. E., & Sutcliffe, K. M. (2007). *Managing the unexpected: Resilient performance in an age of uncertainty*. San Francisco, CA: Jossey-Bass.

CHAPTER 4

MINDING MENTAL HEALTH IN DISASTER RISK REDUCTION AND MANAGEMENT: ENHANCING RESISTANCE THROUGH DISASTER PREVENTION, MITIGATION, AND PREPAREDNESS

Elirozz Carlie Labaria, Avegale Acosta and Charlotte Kendra Gotangco

ABSTRACT

Although the Sendai Framework for Disaster Risk Reduction 2015–2030 explicitly recognizes the need for psychosocial support and mental health services, the focus of this and many disaster risk reduction and management (DRRM) plans lies in the response, recovery, and rehabilitation phases. Less attention has been given to how mental health aspects affect the predisaster phase. This chapter explores the less understood concept of "resistance" in the perspectives model of disaster mental health, which is related to DRRM themes of "prevention and mitigation" and "preparedness" interventions. Four strategies are identified by which DRRM interventions can contribute to psychosocial support and mental health: increasing stress resistance, fostering cohesion and social support, fostering positive cognition, and building self-efficacy and hardiness. We review the cases of the Philippines, Indonesia, Myanmar, and Thailand and report existing socio-political DRRM initiatives for prevention,

Resistance, Resilience, and Recovery from Disasters: Perspectives from Southeast Asia
Community, Environment and Disaster Risk Management, Volume 21, 53–71

ISSN: 2040-7262/doi:10.1108/S2040-726220200000021004

mitigation, and preparedness that can potentially enhance resistance as a pre-disaster intervention. Beyond medical services or clinical mental health interventions for select populations, DRRM interventions can benefit the general public. Despite natural intersections, there remains a need for deliberate and targeted initiatives that explore how vertical pyschosocial care programs can be created to straddle both DRRM and health sectors in practice.

Keywords: Resistance; disaster preparedness; disaster prevention and mitigation; disaster mental health; disaster psychosocial support; sendai framework for disaster risk reduction

Disasters can impact the mental health of the affected population in varying duration, intensity, and expressions (Neria, Nandi, & Galea, 2008; Norris et al., 2002, in Reifels, 2018; Satapathy, 2012; Manzoor & Ali, 2018). According to the World Health Organization (WHO) (2019) epidemiological projections, almost 20–40% of affected populations suffer from mild psychological distress while 30–50% suffers from moderate to severe forms. These impacts manifest as physical, emotional, psychosomatic, cognitive, behavioral, and attitudinal responses and vary from person to person depending on the type of disaster experienced (Satapathy, 2012). In developing countries where most disasters occur, mental trauma is also conceivably high (Manzoor & Ali, 2018). However, literature on disaster-related psychological distress in the Asia, a region particularly vulnerable to natural hazards like earthquakes, volcanic eruptions, typhoons, cyclones, are limited and are more focused on posttraumatic stress disorder resulting from manmade disasters (Kokai, Fuji, Shinfuku, & Edwards, 2004).

The immediate priority of disaster risk reduction and management (DRRM) is usually the physical safety of people and assets, as such, mental health is not keenly attended to. It is not yet fully appreciated how recovery and rehabilitation of people and communities, as well as building their long-term resilience in terms of physical, socioeconomic, and environmental dimensions should be linked to issues of mental health. In addition, the term "mental health" can be misperceived as being overly clinical in approach, prompting international organizations working in the field to coin the term "psychosocial" support to refer to non-medical interventions, but in the process, confusing policy-makers about what approach is needed (WHO, 2007). In addition, the field of mental health and psychosocial support "has been slow to engage with the growing global policy consensus around DRR" (Galappatti, 2015, para. 1).

The Sendai Framework for Disaster Risk Reduction (SFDRR) 2015–2030 (UNISDR, 2015) has shifted the focus from managing the disasters to reducing existing risks and preventing new ones from arising. Covering a wider scope of risks, it features a people-centered, all-hazards, and multi-sectoral approach to risk reduction. Four of the seven SFDRR global targets indicate a strong focus on human health and well-being, moving beyond the aim of simply reducing mortality, and emphasizes resilient health systems. In all four priorities of action, health is underscored through the enhancement of capacities and investments across a spectrum of measures, including safe hospitals and health facilities. The SFDRR also calls for the mainstreaming of disaster risk reduction (DRR) in the laws, regulations, and public policies of the health sector. The actions also extend to strengthening and protection of ecosystem and environmental health and even animal health.

Most notably, the SFDRR includes a provision on disaster mental health (DMH). Under Priority Action 4 on "Enhancing disaster preparedness for effective response" and to "build back better: in recovery, rehabilitation and reconstruction." Section 33 prescribes national and local levels "To enhance recovery schemes to provide psychosocial support and mental health services for all people in need." Although this inclusion is an important recognition in the significant impacts of disasters on mental health, the articulation still suggests a reactionary approach that deals with mental health in the postdisaster phase (response, recovery, and rehabilitation).

Moreover, operationalizing this global call involves the challenges of articulating, linking, and reconciling the conceptual underpinnings of DRRM and DMH as two distinct fields of practice; something that the SFDRR does not provide elaboration and guidance on. Reifels (2018, p. 2) stated this problem as "a lack of clarity about the conceptual interlinkages and place of DRR principles within the field of DMH, which is traditionally marked by a prevailing recovery orientation." Without tackling this challenge, developing strategies to reduce and manage mental health risks and impacts might disregard or overlook potential interventions in the predisaster phase.

This review aims to contribute to the elaboration of the conceptual linkages of DRRM and DMH by investigating the intersection of the leading frameworks. It proposes a consolidated framework for disaster management (DM), DRR and disaster risk management (DRM), referred to here as DRRM, based on the United Nations International Strategy on Disaster Reduction (UNISDR) (2007) and the Johns Hopkins Perspectives Model of Disaster Mental Health (henceforth, perspectives model) (Kaminsky, Lee McCabe, Langlieb, & Everly, 2006). The DRRM framework which has been widely used in science-policy practice deals with the thematic areas of work along the phases of the DM cycle: prevention and mitigation, preparedness, response, and recovery and rehabilitation. The Perspectives Model is formulated around the notions of resistance, resilience, and recovery. This consolidated framework is intended to help bridge the gap between researchers and practitioners of the DMH and DRRM fields, so that DMH issues can better be integrated and mainstreamed into prevailing DRRM frameworks and policies, and vice versa.

This review particularly explores the less understood DMH concept of "resistance" which is deeply tied to DRRM themes of "prevention and mitigation" and "preparedness" interventions involved in the *ex ante* phases of the cycle. It may be that DRRM interventions that build resistance increase the chances of survival, effectively reducing mortality, potentially extending to those who are dependent on the survivor. Furthermore, building resistance is proposed to be crucial in fostering long-term resilience and determining the quality and speed of recovery and rehabilitation of people and communities. It can also reduce mental health problems caused by traumatic events such as posttraumatic stress disorder, depression, anxiety, grief, and substance abuse. Thus, this review examines criteria by which DRRM policies, plans, and initiatives contribute toward building mental health resistance, and applies the proposed criteria to a scoping of existing DRRM measures in the Philippines and selected Southeast Asian countries. This scoping will help identify potential starting points for a more intentional integration of DMH and DRRM.

CONCEPTUAL LINKAGES OF DRRM AND DMH

Unpacking the language relating to disaster risk governance is a vital step toward locating the conceptual linkages. The UNISDR defines key terms as follows (UNISDR, 2007):

Disaster Management

The organization and management of resources and responsibilities for dealing with all aspects of emergencies, particularly preparedness, response, and rehabilitation. It is interchangeable with the term emergency management.

Disaster Risk Reduction

The conceptual framework of elements considered with the possibilities to minimize vulnerabilities and disaster risks throughout a society, to avoid (prevention) or to limit (mitigation and preparedness) adverse impact of hazards, within the broad context of sustainable development.

Disaster Risk Management

The systematic process of using administrative decisions, organization, operational skills, and capacities to implement policies, strategies, and coping capacities of the society and communities to lessen the impacts of natural hazards and related environmental and technological disasters. This comprises all forms of activities including structural and non-structural measures to avoid (prevention) or to limit (mitigation and preparedness) the adverse effects of hazards within the broad context of sustainable development.

A recent UN intergovernmental expert working group provided recommendations for terminology for DRR. DRR is the policy objective of DRM and its goals and objectives are defined in DRR strategies and plans, while DRM is the application of DRR policies and strategies (United Nations General Assembly (UNGA), 2016). Meanwhile, DM (interchangeably, emergency management) is the organization, planning, and application of measures as focusing on the *post ante* phases of response, recovery, and rehabilitation, although preparedness remains a key element.

Although these varied definitions have guided nations in crafting and updating their respective legislative frameworks and institutions for disasters, a lack of appreciation for the overall dynamics of risk as part of both short- and long-term development may encourage overemphasis on one aspect of the cycle (e.g., only on the *post ante* phases) and side-track the critical relationship among them. This study thus uses the term **DRRM** to refer to the whole DM cycle from prevention and mitigation, preparedness, response, recovery, and rehabilitation, and thus contains all the elements described in DM, DRR, and DRM. The value of using an encompassing concept is to better appreciate the continuity and linkage of actions. This is especially useful in strategic planning to ensure that interventions that cross-cut several phases of the management cycle are properly considered and any ambiguity in the governance of such are addressed.

The Perspectives Model of DMH developed at the Johns Hopkins University (Kaminsky et al., 2006) is a strategic and integrative paradigm that considers multiple interventions to be applied in a tactical sequence in order to create the most effective approach to mental health in a disaster incident. This evidence-based, outcome-driven model moves away from reactionary nature of prevailing DMH practices. Current practices mostly focus on providing interventions for those who were already affected, and clinical interventions for those who were traumatized by the disaster. The perspectives model goes beyond providing interventions during and postdisaster. It opens the discourse into interventions that support DMH, which includes psychosocial interventions before disaster strikes (Kaminsky et al, 2006). This can help increase the ability of the individual and the community to withstand the effects of the disaster and develop self-efficacy on what to do when it comes.

Kaminsky et al. (2006) provides succinct definition of the three concepts engaged in the model. *Resistance* refers to the ability of an individual, a group, an organization, or even an entire population to literally resist manifestations of clinical distress, impairment, or dysfunction associated with critical incidents, terrorism, and even mass disasters. Resistance is likened to psychological or behavioral immunity to distress and dysfunction. *Resilience* refers to the ability of an individual, a group, an organization, or even an entire population to rapidly and effectively rebound from psychological and/or behavioral perturbations associated with critical incidents, terrorism, and even mass disasters. *Recovery* refers to the ability of an individual, a group, an organization, or even an entire population to literally recover the ability to adaptively function, both psychologically and behaviorally, in the wake of a significant clinical distress, impairment, or dysfunction subsequent to critical incidents, terrorism, and even mass disasters.

Note that these three concepts follow the temporal trajectory of human perception and reaction to disaster situations. However, they do not operate in isolation – resistance is closely linked to resilience. Given that resilience pertains to the bouncing back from psychological trauma, stress that is not successfully resisted may impair this ability to rebound and affect the quality of recovery.

The psychosocial lens of the Perspectives Model allows it to be contrasted to DRRM. Fig. 1 locates the model in DRRM framework. Building resistance is akin to "psychological immunization" and is considered a predisaster intervention. It is aligned with actions for disaster prevention and mitigation and disaster

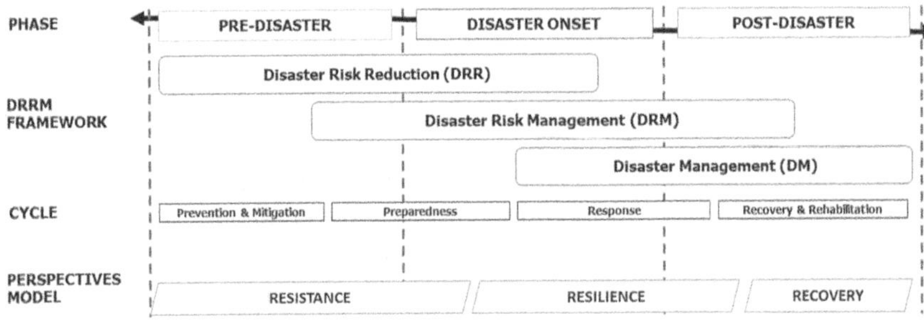

Fig. 1. Locating DMH Perspectives Model in the DRRM Framework.

preparedness. Mental health interventions during the resistance phase can also be used as a tool and strategy for improving preparedness and for enhanced protection of lives and property, and improved safety (Roudini, Khankeh, & Wiltruk, 2017). Because enhancing resistance is closely linked with resilience, the former may also contribute toward actions for better response and recovery.

STRATEGIES BY WHICH DRRM CAN CONTRIBUTE TO MENTAL HEALTH RESISTANCE

Resistance refers to protective factors that mitigate psychological distress. In their discussion of the perspectives model, Kaminsky et al. (2006) proposed four strategies to increase resistance to psychological distress in times of disasters: (a) increasing stress resistance, (b) fostering cohesion and social support, (c) fostering positive cognition, and (d) building self-efficacy and hardiness. Here, we review these strategies and how they may be addressed through DRRM initiatives.

Increasing Stress Resistance

Stress resistance may refer to structures set in place to help individuals and communities to be mentally prepared and to minimize psychological distress due to disasters. Disaster prevention and mitigation and disaster preparedness not only reduce loss of life and property, but also feelings of fear and anxiety (Morrissey & Reser, 2003). These include planning strategies such as developing organizational support systems; conducting needs assessments; training of first responders; creating early warning systems and evacuation plans; and geographical mapping of existing resources, risk areas, and vulnerable groups and translating such information to knowledge that can be utilized by the community (Reifels, 2018). The Inter-agency Standing Committee on Mental Health and Psychosocial Support in Emergency (IASC, 2007) also made a comprehensive list of strategies that can be used as responses during the emergency phase of disaster (resilience) and can also be applied to early reconstruction efforts (recovery). Later on, some of these efforts can be institutionalized to increase community resistance. These may include: establishing sustainable coordination structures which include the government and stakeholders from the community, integrating MHPSS activities into the development of national policies and programs, developing of emergency preparedness plans for institutions, strengthening the community's ability to protect human rights, planning steps to immediately provide physical safety for disaster survivors, and providing support to first responders during emergencies. It is also crucial during the predisaster phase to train and supervise mental health workers and volunteers who may be deployed to provide psychosocial response in times of crisis and emergencies. These trainings include psychological first aid (PFA), gender sensitivity, reducing discrimination and stigma for people with mental illness, and prevent and address other issues such as substance use.

Organizations and communities may also help individuals develop resilience to stressful events by helping them to develop stress management and coping strategies that can increase their ability to recover after a traumatic event. These may include relaxation techniques that can help manage physiological arousal, adaptive coping strategies such as identifying strengths, managing thoughts and emotions, problem-solving, seeking social support, promoting positive activities, and planning for the future (Hechanova et al., 2018; Lating et al., 2003). First responders are also vulnerable to experiencing secondary trauma, and are at a high risk for traumatic stress reactions, thus they may benefit from resilience trainings to further equip them in handling stressful events (Lowry & Lating, 2002).

Fostering Cohesion and Social Support

Cohesion and social support have been found to be a protective factor when facing psychological distress. In a study conducted among earthquake survivors and flood-affected communities, it was found that participation in a 3-day integrated disaster preparedness interventions decreased depression and posttraumatic stress symptoms (PTSD) symptoms, and increased social cohesion (Welton-Mitchell, James, Khanal, & James, 2018). The disaster preparedness intervention utilized an integrated and experiential approach which involved safe spaces for exchange of personal experiences and support. The intervention also included ways to practice coping for disaster-related stress, and training on how to respond and how to support oneself and others when a disaster strikes. The intervention was found to aid in increasing social cohesion among those who were part of the intervention group. There was also an increase in intent to provide help and support for disaster and mental health related concerns, and an increase in help-seeking behaviors for disaster-related concerns among the members of the community (James, Welton-Mitchell, Noel, & James, 2019). Communities that have higher levels of community resilience tend to recover quicker than those with low social cohesion. When an individual perceives that their community is prepared for disasters, and have high degree of assistance from other members, they are more likely to engage in disaster preparedness behaviors (Espina & Teng-Calleja, 2015; Sagala & Okada, 2009). On an organizational level, a study by McCabe et al. (2014) discussed how a partnership among academicians, government agencies, and faith-based groups in developing a disaster preparedness intervention composed of PFA and guided preparedness planning can help increase public health preparedness and community resilience. The provision of recreational and cultural spaces was also known to help increase cohesion in the community (Lowry & Lating, 2002).

Disasters increase the risk of relocation from homes and separation from families. This separation exposes individuals to added psychological vulnerabilities. As much as possible, evacuation plans must prioritize keeping families together to minimize psychological distress (Jacob, Mawson, Payton, & Guignard, 2008). A study among children survivors of disaster found that those who were separated from parents showed more signs of emotional stress as opposed to those who stayed with at least one parent (Wolff & Fesseha, 2005).

Fostering Positive Cognition

One's cognitions and appraisals about hazards are correlated with one's tendency to prepare for it. Mishra and Suar (2007) emphasized the importance of disaster education as it seems to predict risk perception. Risk perception is one of the predictors of anticipatory behaviors toward disasters (Espina & Teng-Calleja, 2015). How people are educated about disasters and how media illustrates hazards pertaining to it may influence risk perception. If people perceive that they can do something to protect themselves when a hazard occurs, there is a higher likelihood that they may act on it. In addition, lack of knowledge and uncertainty about one's ability to handle certain hazards may lead to apathy, which then reduces disaster preparedness behaviors (Johnston et al., 2005).

Culture, values, and beliefs may also influence cognition and appraisal when it comes to disaster preparedness. Attributions about hardships can influence how people handle crisis and difficulties they face. For instance, ascribing religious significance to hardships (e.g., "God has a plan"), seeking peace of mind, and collaborating with God (e.g., "When you have God everything becomes light"), helps individuals reframe their hardships and view it in a more positive light (Nakonz & Shik, 2009)

Building Self-efficacy and Hardiness

Self-efficacy refers to the belief in one's ability to accomplish certain tasks or to succeed in certain situations. Bandura postulated that self-efficacy can be developed through successful experiences in doing that task, observing others as they succeed, through encouragement, and support, and through one's ability to manage emotional and physical states (Bandura, 1994). Simulations of disaster preparedness have been known to increase one's efficacy in responding to disaster situations and lowers fears and anxiety about the disaster (Galappatti & Richardson, 2016; Juanita, Suratmi, & Maghfiroh, 2018).

Given these four strategies for increasing mental health resistance, the scoping of DRRM literature reflects that disaster prevention and mitigation and preparedness initiatives can already contribute toward increasing the individual and the community's ability to withstand psychological distress during times of

Table 1. Potential Intersections Among Two of the Thematic Areas of the DRRM Cycle and the Strategies Contributing to Mental Resistance.

DRRM Thematic Area	Activities	Strategies that Contribute to Mental Resistance			
		Increasing Stress Resistance	Fostering cohesion and social support	Fostering positive cognition	Building self-efficacy and hardiness
Disaster Prevention and Mitigation	Conduct of local risk assessments (include hazards evaluation, exposure and vulnerability analyses, mapping and planning exercises)	✓			✓
	Establishment of early warning systems and corresponding response/evacuation plans	✓		✓	✓
	Community-based mapping and capacity-building		✓		
Disaster Preparedness	Community preparedness training (e.g., evacuation drills, hazard simulation exercises, response trainings)	✓	✓	✓	✓
	Information, education and communication; activities for raising awareness and encouraging public participation	✓	✓	✓	✓

disaster. Table 1 lists common activities in the relevant DRRM thematic areas of prevention and mitigation, and preparedness, and indicates the strategies they potentially employ that increase mental health resistance. The next section discusses how these strategies are incorporated in specific DRRM programs in the Philippines and Southeast Asia.

CASE STUDIES IN SOUTHEAST ASIA

The Philippines

Although Republic Act 10121 (2010) on strengthening the Philippine DRRM system does not contain explicit reference to mental health or psychosocial support, the Philippines DRRM Plan for 2011–2028 (NDRRMC, 2011) includes two outcomes directly related to mental health (summarized on page 4). The first falls under the thematic area of Disaster Response, seeking "psychosocial needs of directly and indirectly affected population addressed" led by the country's Department of Health (DOH). Activities include coordination among heads of MHPSS, psychosocial programs and referrals, and conduct of traumatic and/or psychological stress debriefings. The second outcome, co-led by the DOH and the Department of Social Welfare and Development, is under the Disaster Rehabilitation and Recovery theme, which targets that "a psychologically sound, safe and secure citizenry that is protected from the effects of disasters is able to restore to normal functioning after each disaster." Activities largely pertain to risk protection measures; one specifically seeks to "build capacities of psychosocial care providers."

The National Disaster Response Plan for Hydro-meteorological Hazards (NDRRMC, 2014) operationalized a Health Cluster led by the DOH which includes a Mental Health and Psychosocial Support (MHPSS) sub-cluster, along with the sub-clusters on public health, nutrition, sanitation and hygiene of the DOH–Health Emergency Management System, to provide support for the delivery of "appropriate and timely health services to the affected population" (NDRRMC, 2014, p. 65). The MHPSS sub-cluster particularly aims to (1) prevent and reduce prevalence of negative mental health and psychosocial consequences of emergencies and disasters and (2) to enhance community resiliency to psychosocial impacts of emergencies and disasters. The Philippines has decentralized public health system managed at the lowest level by Local Government Units (LGUs) and overseen in the national level by the Department of Health (DOH) (DOH, Republic of the Philippines, 2012).

DOH through the Health and Emergency Management Staff has produced a Pocket Emergency Tool (PET) in collaboration with the WHO to serve as reference materials for technical staff on pointers and technical guidelines for emergency management from preparedness to response to recovery. In the phasing for emergency response operations, addressing psychosocial and mental health concerns are prescribed as extended response from 12 to 24 h after an emergency situation. The Integrated Code Alert System adopted by the health sector prescribes mental health professions to be available at DOH hospitals around the country when a Code Blue early warning alert (the second highest alert, requiring 50% of all hospital personnel to report for duty) for an impending threat or incident is activated (DOH, 2012).

More saliently, the PET sets out domains and minimum responses in emergencies for MHPSS. It details the common functions on multi-sectoral coordination, initiation of participatory systems for assessment, monitoring, and evaluation, and application of protection and human rights standards to be followed by all humanitarian actors, as well as the necessary capacity building on Basic MHPSS, PFA, and Stress Management for human resources such as staff and volunteers. It further outlines core mental health and psychosocial support to include (1) facilitating community mobilization and support, (2) ensuring provision and access to psychological interventions and mental health treatment for especially for those with severe mental disorders, (3) restoring access to normal to education or provide equivalent informal education, and (4) dissemination of information to the affected population on relief efforts, their legal rights, and positive coping methods. Social considerations in terms of food security and nutrition, shelter and site planning, as well as water and sanitation are also emphasized as essential. Most importantly, the tool elaborates on the steps of MHPSS interventions from assessment to provision of PFA, psychosocial processing, early social interventions for children and families, and indications of psychosocial concerns for disaster workers (DOH, 2012).

Interventions that are explicitly articulated as addressing mental health are evidently and unsurprisingly geared toward the response and recovery phases and tied to the operations of the public health sector system. They are thus responsive to resilience and recovery dimensions of the Perspectives Model. Investigating resistance entails looking at the other aspects of DRRM that are not explicitly on health and thus do not require medical/clinical actions.

In the NDRRM Plan, for instance, prevention and mitigation and preparedness interventions may contribute to building resistance in the very nature of the activity. The law requires local risk assessments including hazards evaluation, exposure and vulnerability analyses, mapping and planning exercises. Especially when conducted in a community-based, participatory manner, these assessments enhance the understanding of people and communities of their risks for them to be mentally prepared for potential impacts and gives them the opportunity to reduce their own risks and prepare for adaptive or coping strategies. These measures inherently contribute to the strategies to build mental health resistance. The community-based approach fosters social cohesion and support. In addition, mental preparedness can increase stress resistance and the opportunities to build self-efficacy and hardiness.

The establishment of early warning systems, the development and dissemination of information, education, and communication materials, awareness raising campaigns, and the conduct of preparedness trainings (such as evacuation drills, hazard simulation exercises, response trainings) help to condition communities to anticipate the risks and be responsive, thus increasing stress resistance, fostering positive cognition and building self-efficacy and hardiness. These measures structure the cues from which individuals can initiate the risk-reduction actions available to them. They also give people a sense of active participation in a public intervention, potentially reducing feelings of isolation during stressful events and thus fostering cohesion and social support.

Meanwhile, the integration of DRRM in school curricula is a model of a long term, inter-generational approach toward building resistance. The Department of Education Order No. 37, s. 2015, proposes the Comprehensive Disaster Risk Reduction and Management in Education Framework which also adheres to the four thematic areas of the DRRM. It is concerned not just with the content of the curriculum but also the conditions of its delivery. Section V.3 explicitly recognizes that the psychosocial conditions of both the students and the teachers affect the quality of the education.

Building the capacities of local governments in terms of the development of local DRRM policies, plans, and budgets, and skill trainings for DRRM workers helps increase their perceived credibility in the eyes of their constituent communities, potentially increasing responsiveness. The NDRRM Plan also emphasizes "strengthened partnership and coordination and among all key players and stakeholders" (Outcome 11) in DRRM which include civil society organizations and the private sector also further strengthens a sense of community and support system among those affected (Fostering cohesion and social support). It is important to note, however, that building capacities of local governments and other stakeholders in leadership positions should extend beyond the application of template materials or methods, to the contextualization of educational and advocacy materials. Emerging research indicates that culture influences the understanding and perception of risk – local values and beliefs may influence receptiveness to new knowledge and actions taken in response (Gotangco & de Leon, 2018).

Outside of the public health sector system and the NDRRMP, other organizations in the Philippines are spearheading initiatives to enhance resistance by helping build self-efficacy of different stakeholder groups. For instance, a local channel partnered with the Psychological Association of the Philippines (PAP) to create videos that demonstrates how PFA can be provided in times of crises and emergencies. The PAP also developed a six-module resilience intervention based on the needs of disaster survivors. The program was found to be effective in decreasing PTSD, anxiety, and depressive symptoms, and increasing coping self-efficacy of survivors (Hechanova et al., 2015; Hechanova, Waelde, & Ramos, 2016). The program was designed to be simple enough so that trained lay facilitators could implement the program (Hechanova et al., 2018). In addition, UNESCO and the Government of Japan provided the Philippines' Department of Education with a manual that aims to equip secondary school teachers with skills during disasters and emergency situations. This includes skills on PFA and in implementing a resilience intervention that also includes a disaster preparedness component into it (UNESCO, 2015). At the tertiary level, a university in the Philippines has offered a course on disaster and its effect on mental health.

Myanmar and Thailand

Other countries in the Southeast Asian region show similarities to the Philippine case. For example, the Myanmar National Disaster Management Law of 2013 and the Thailand Disaster Prevention and Mitigation Act of 2007 have no explicit mention of the terms "mental health" or "psychosocial support." Lack of explicit articulation notwithstanding, the corresponding action plans do recognize mental health and psychosocial needs, however, the "disaster preparedness" aspect is discussed more in the context of building capacities for better emergency response, recovery, and rehabilitation. For example, the Thailand National Disaster Risk Management Plan (2015) delegates to mental health programs to the Ministry of Public Health, and psychological rehabilitation through social interactions to the Ministry of Social Development and Human Security, as part of the emergency support functions for which they need to plan. In the Myanmar Action Plan on Disaster Risk Reduction (NDMC, 2017a), under Pillar 4 on *Enhancing disaster preparedness for effective response and resilient rehabilitation and reconstruction*, the Ministry for Health and Sports is mandated to oversee health sector preparedness for response and recovery, which includes psychosocial support.

Indonesia

In Indonesia, the Law of the Republic of Indonesia of 2007 concerning DM (State Gazette of Republic of Indonesia Number 66, 2007) does explicitly state that disasters have psychological impact and psychosocial support is a basic necessity. These are recognized as components of postdisaster rehabilitation – again, the recognition of mental health and psychosocial needs is more in the context of the recovery and rehabilitation phase. The Disaster Risk Management Baseline Status Report (Badan Nasional Penanggulangan Bencana (BNPB)/ National Disaster Management Authory, 2015) also categorizes psychosocial needs under the Humanitarian aspects of postdisaster support. Among the issues identified for action is the strengthening of psychosocial services particularly for vulnerable groups during the period of disaster response, relief, and transition – also more in the context of emergence services rather than the preparedness or prevention and mitigation phases. However, there do exist provisions in the national law on the rights of the community which includes "[participating] in planning, operation, and maintenance of healthcare aid program including psychosocial support" (Article 26). The planning process is the closest aspect relating to the preparedness phase.

SUMMARY

A review of the disaster preparedness measures instituted by these countries reveals that, like the Philippines, there are plans and actions that are not deliberate nor targeted mental health initiatives but nevertheless contribute to enhancing mental health. These are the standard measures of improving on early warning systems, information dissemination reaching the farthest populations, disaster awareness campaigns, and simulation exercises or drills. These all serve to build stress resistance by enabling realistic preparations and providing opportunities for realistic pre-incident training, and to improve on group cohesion by refining communication networks and protocols among concerned agencies and stakeholders. One example of a disaster literacy measure for the general public is the International Tsunami Museum established in Thailand, where people can come and learn about tsunamis, its warning systems, as well as appropriate responses when a tsunami happens. For those who did not have access to museums, educational videos were provided through schools for free. Evaluations found both strategies to increase knowledge of disasters and appropriate response and decrease levels of fear (Sattler, 2008). Efforts toward community-based disaster resilience – such as the development of the 2017 Myanmar National Framework for Community Disaster Resilience (NDMC, 2017b), or the formation of Kelompok Siaga Bencana (Disaster Community Groups) in Indonesia (Rahayu & Ophiyandri, 2018) – contribute toward a sense of connectedness, direction, and motivation within the community, which all lead to fostering group cohesion and social support that can help buffer stress. Drawing from volunteerism, as advocated by both the Myanmar action plan and community resilience framework, can also help build self-efficacy among those involved.

In terms of policies and plans targeting mental health issues specifically, Indonesia, Thailand, Malaysia, Myanmar, Mongolia, and Singapore are among the countries with existing mental health legislation and policies or plans (Ito, Setoya, & Suzuki, 2012). Indonesia and Thailand, two countries worst affected by the 2004 tsunami, are among those advocating for integration of mental health into primary health care, and subsequently taking initiative to better prepare medical professionals for emergency mental health and psychosocial services (Panyayong & Pengjuntr, 2006; Prasetiyawan, Viora, Maramis, & Keliat, 2006; Setiawan & Viora, 2006). Thailand's health policies are part of the National Health Act of 2007, and a Mental Health Act was passed in 2008. Similarly, Indonesia's mental health policies are integrated in Health Law number 36/2009 (WHO, 2011), and a Mental Health Law was passed in 2014 (Yusuf, 2016). In addition, there have been proposals in Indonesia to include disaster preparedness as part of the nursing curriculum, as initial evaluations on simulation classes proved to increase efficacy to respond during emergencies as opposed to those who only have the theoretical knowledge for it (Juanita et al., 2018).

Setiawan and Viora (2006) further describe the principles by which a DMH preparedness plan was being developed in Indonesia. These principles include contingency planning and implementation of assessments prior to interventions, training and supervision for providers of PFA, collaboration with other relevant agencies, monitoring outcome indicators, and adopting a long-term perspective. Initiatives guided by such principles would not only build competency and efficacy in medical professionals but would also translate toward helping to buffer stress in the communities they serve. However, whether the implementation of such initiatives is successful is a different matter. A study in Aceh by Ismail, Suwannapong, Howteerakul, Tipayamongkholgul, and Apinuntavech (2016) of the Community Mental Health Nurse Program, which started in 2006, showed that the program had not yet reached its target outcomes. The results of the community-based descriptive household survey found that the community-education intervention was ineffective, and suggested that this could be due to lack of understanding of the local culture of Aceh (i.e., belief that tsunami is a "punishment from God," and the use of religion to deal with pain and loss). Similar to the Philippine experience, this indicates that effectiveness is best achieved through contextualization within the specific culture of a community.

DISCUSSION AND RECOMMENDATIONS

Clearly, despite the lack of comprehensive – or even explicit – mental health and psychosocial support initiatives, there are natural synergies between DRRM and DMH which can be harnessed in a more deliberate way to efficiently address the objectives of the two fields of practice. As shown in the review, common initiatives for disaster prevention and mitigation and disaster preparedness already contribute to increasing stress resistance, fostering cohesion and social support, and building self-efficacy. As the proposed consolidated framework shows, enhancing resistance as a predisaster intervention intersects with DRRM actions for prevention and mitigation. Preparedness potentially benefits the general public and may reduce the need for medical or clinical mental health services. Thus, there is impetus to be more attentive to what DRRM can do, create, or respond to outside of clinical mental health interventions, and to explore how psychosocial care programs, coordinated both vertically and horizontally across relevant government and other institutions, can be created to straddle both DRRM and health sectors in practice. In addition, a caveat of the review conducted here is that discrepancies may be observed between existing policies and literature and implementation on the ground. This is an unfortunate but common gap, which necessitates closer monitoring and evaluation to ensure that DRRM initiatives correctly and adequately employ mental health resistance strategies.

Using a DRRM perspective, a basic starting point is the improved implementation of prevention and mitigation and disaster preparedness tasks across all the cities and municipalities. Manifestations of resistance-building initiatives in the local levels already exist; however, they are not consistent across all localities, and are usually appreciated only as procedural components of DRRM. Improving these interventions would not only mean making them technically sound but also recalibrating them in such a way that they are responsive to the mental health dimensions of the human person. For instance, a key need is the generation, validation, and updating of high-resolution hazard, exposure, vulnerability and risk maps, which serve as the bases for preparedness planning and contributed toward increasing stress resistance. Not all cities and municipalities in the Philippines have these resources or capacities, and the case is similar for other Southeast Asian countries. In areas where technology is limited, or in cases when national government or external agencies generate these maps, community-based participatory mapping methods, community-based approaches can be employed for local validation, given its co-benefits for improving on the accuracy of these knowledge products and for fostering social cohesion.

Developing effective mechanisms for DMH and DRRM therefore requires contextualization – that is, consideration of local conditions and cultures in the design and implementation of these mechanisms. For example, community-based early warning systems could be crafted in locally or culturally preferred and appropriate ways. If warning devices like sirens and blare horns cause panic to some groups, then alternative approaches such as telephone trees or even door-to-door warnings can be used. Implementing drills in communities has also been a common practice, and these can be enhanced to be more comprehensive and situational, that is, to simulate different scenarios in the specific community and build citizens' positive cognition and self-efficacy in terms of making decisions on appropriate actions as different circumstances arise, and not just in terms of following pre-determined instructions. The planning of evacuation centers and camp management during the crisis can be done in such a way that families, clans, or even community groups are able to stay together and plan response and recovery actions together. Coordination with other stakeholders including social clubs and religious groups may further foster social support in communities where these actors play an influential role in community dynamics.

WHO (2007) compiled a list of essential components for a disaster preparedness plan to address the mental health and psychosocial aspects. A few items on this list are already components of any standard DRRM plan, such as the development of a risk communication strategy; the establishment of up-to-date and well-maintained communication equipment and infrastructure; and empowerment at the local level to launch first response actions. Other items would likely exist in some form in a DRRM plan but need to be tweaked for mental health and psychosocial care considerations, specifically. These easily implemented actions include identifying responsible persons and agencies in the coordination mechanisms; articulating the roles and responsibilities of community-level workers, and developing guidelines for how they should operate in a community given local cultures.

The remaining items on the WHO list represent more deliberate and targeted interventions from the field of DMH. These include the allotment of resources specifically for DMH and psychosocial programs in disaster preparedness plans; the development, dissemination, and validation of technical materials for different stakeholders (e.g., training materials and guidelines for DMH professionals as well as the general public); the design of a plan for assessing psychosocial distress, including the appropriate instruments for use; and the development of a long-term community health system. In addition to these interventions for DMH specifically, initiatives that aim to build life skills in general (e.g., in the form of leadership and personality or skills trainings and seminars, in contrast to awareness campaigns and formal education) can increase positive cognition in a wider scale, thus should be well designed and strategically implemented. All these initiatives may require more time and attention to put in place, but are well worth it in order to achieve the synergy between DRRM and DMH.

ACKNOWLEDGMENTS

This work was partially supported by the Capacity Building in Asia for Resilience Education (CABARET) project co-funded by the Erasmus+ Programme of the European Union.

REFERENCES

Badan Nasional Penanggulangan Bencana (BNBP)/National Disaster Management Authory. (2015). *Indonesia's disaster risk management baseline status report 2015*. Retrieved from https://www.preventionweb.net/publications/view/50832

Bandura, A. (1994). Self-efficacy. In V. S. Ramachaudran (Ed.), *Encyclopedia of human behavior* (Vol. 4, pp. 71–81). New York, NY: Academic Press. (Reprinted in H. Friedman [Ed.], *Encyclopedia of mental health*. San Diego, CA: Academic Press, 1998).

Department of Education (DepEd), Republic of the Philippines. (2015). *The Comprehensive Disaster Risk Reduction and Management in Education Framework*. DepEd Order No. 37, s. 2015. Retrieved from http://depedmabalacat.org/announcements/93-comprehensive-disaster-risk-reduction-and-management-drrm-in-education-framework

Department of Health (DOH), Republic of the Philippines. (2012). *Pocket emergency tool* (4th ed.). Department of Health (DOH), Republic of the Philippines, in collaboration with the Emergency and Humanitarian Action unit, Regional Office for the Western Pacific, World Health Organization.. Retrieved from http://ro11.doh.gov.ph/pdf/PET.pdf

Espina, E., & Teng-Calleja, M. (2015). A social cognitive approach to disaster preparedness. *Philippine Journal of Psychology*, *48*(2), 161–174.

Galappatti, A. (2015). Better late than never: Integrating mental health and psychosocial support into disaster risk reduction. Retrieved from http://wp.preventionweb.net/wcdrr/better-late-thannever-integrating-mental-health-psychosocial-support-into-disaster-risk-reduction/

Galappatti, A., & Richardson, S. M. (2016). Linking mental health and psychosocial support and disaster risk reduction: Applying a wellbeing lens to disaster risk reduction. Retrieved from https://www.interventionjournal.com/sites/default/files/Linking_mental_health_and_psychosocial_support_and.6.pdf

Gotangco, C. K., & de Leon, I. P. (2018). Understanding communication needs: A Marikina barangay experience linking flooding to climate change communication. In *Handbook of climate change communication* (Vol. 3, pp. 85–98). Cham: Springer.

Hechanova, M. R., Docena, P. S., Alampay, L. P., Acosta, A. C., Porio, E. E., … Berger, R. (2018). Evaluation of a resilience intervention for Filipino displaced survivors of Super Typhoon Haiyan. *Disaster Prevention and Management*, *27* (3), 346–399. doi:10.1108/DPM-01-2018-0001

Hechanova, M. R., Waelde, L. C., Docena, P. S., Alampay, L. P., Alianan, A. S., Flores, M. J., Melgar, I. E. (2015). The development and initial evaluation of Katatagan: A resilience intervention for Filipino disaster survivors. *Philippine Journal of Psychology*, *48*(2), 105–131.

Hechanova, M. R., Waelde, L. C., & Ramos, P. A. (2016). Evaluation of a group-based resilience intervention for typhoon Haiyan Survivors. *Journal of Pacific Rim Psychology*, *10*(12). https://doi.org/10.1017/prp.2016.9

Inter-agency Standing Committee. (2007). IASC guidelines on mental health and psychosocial support in emergency settings. Retrieved from https://www.refworld.org/docid/4e97ceca2.html

Ismail, N., Suwannapong, N., Howteerakul, N., Tipayamongkholgul, M., & Apinuntavech, S. (2016). Assessing disaster preparedness and mental health of community members in Aceh, Indonesia: A community-based, descriptive household survey of a national program. *Rural & Remote Health*, *16*(4).

Ito, H., Setoya, Y., & Suzuki, Y. (2012). Lessons learned in developing community mental health care in East and South East Asia. *World Psychiatry*, *11*(3), 186–190.

Jacob, B., Mawson, A. R., Payton, M., & Guignard, J. C. (2008, September–October). Disaster mythology and fact: Hurricane Katrina and social attachment. *Public Health Reports*, *123*, 555–566.

James, L. E., Welton-Mitchell, C., Noel, J. R., & James A. S. (2019). Integrating mental health and disaster preparedness in intervention: A randomized controlled trial with earthquake and flood-affected communities in Haiti. *Psychological Medicine*, 1–11. https://doi.org/10.1017/S0033291719000163

Johnston, D., Paton, D., Crawford, G. L., Ronan, K., Houghton, B.,& and Burgel, P. (2005). Measuring tsunami preparedness in coastal Washington, United States. *Natural Hazards*, *35*, 173–184.

Juanita, F., Suratmi, S., & Maghfiroh, I. L. (2018). The effectiveness of basic training on disaster management pilot program for disaster preparedness in community. *Indonesian Nursing Journal of Education and Clinic*, *2*(2), 126–135. doi:10.24990/injec.v2i2.157

Kaminsky, M., Lee McCabe, O., Langlieb, A., & Everly, G. (2006). An evidence-informed model of human resistance, resilience, and recovery: The Johns Hopkins' Outcome-Driven Paradigm for Disaster Mental Health Services. *Brief Treatment and Crisis Intervention*, *7*(1), 1–11.

Kokai, M., Fuji, S., Shinfuku, N., & Edwards, G. (2004). Natural disaster and mental health in Asia. *Psychiatry and Clinical Neurosciences*, *58*(2), 110–116.

Lating, J. M., Everly, G. S., Peragine, T. F., Neel, M. L., Glock, N. P., & Sherman, M. F. (2003). Biofeedback-assisted relaxation as part of preincident stress-management training within a model of comprehensive crisis intervention: A pilot study. *Brief Treatment and Crisis Intervention*, *3*(4), 437–443.

Lowry, J. L., & Lating, J. M. (2002). Reflections on the response to mass terrorist attacks: An elaboration on Everly and Mitchell's 10 commandments. *Brief Treatment and Crisis Intervention*, *2*(1), 95–104.

Manzoor, S., & Ali, M. (2018). Disaster mental health: A need for multipronged approach. *Indian Journal of Health and Well-Being*, *9*(3), 483–488.

McCabe, O. L., Semon, N. L., Lating, J. M., Everly, G. S., Jr, Perry, C. J., Moore, S. S., … Links, J. M. (2014) An academic-government-faith partnership to build disaster mental health preparedness and community resilience. *Public Health Reports*, *129* (Supplement 4), 96–106.

Mishra, S., & Suar, D. (2007). Do lessons people learn determine disaster cognition and preparedness. *Psychology and Developing Societies*, *19*(2), 143–159. doi:10.1177/097133360701900201

Morrissey, S. A., & Reser, J. P. (2003). Evaluating the effectiveness of psychological preparedness advice in community cyclone preparedness materials. *The Australian Journal of Emergency Management*, *18*(2), 46–61.

Myanmar National Disaster Management Law: Pyidaungsu Hluttaw Law/No. 2. (2013). Retrieved from http://www.myanmar-law-library.org/law-library/laws-and-regulations/laws/myanmar-laws-1988-until-now/union-solidarity-and-development-party-laws-2012-2016/myanmar-laws-2013/pyidaungsu-hluttaw-law-no-21-2013-natural-disaster-management-law-burmese-and.html

Nakonz, J., & Shik, A. W. Y. (2009). And all your problems are gone: Religious coping strategies among Philippine migrant workers in Hong Kong. *Mental Health, Religion & Culture*, *12*(1), 25–38. doi:10.1080/13674670802105252

National Disaster Management Committee (NDMC), Republic of the Union of Myanmar. (2017a). Myanmar action plan on disaster risk reduction. Retrieved from http://www.mm.undp.org/content/myanmar/en/home/library/environment_energy/Myanmar_Action_Plan_DRR_2017.html

National Disaster Management Committee (NDMC), Republic of the Union of Myanmar. (2017b). Myanmar national framework for community disaster resilience. Retrieved from https://www.preventionweb.net/english/professional/policies/v.php?id=52573

National Disaster Risk Reduction and Management Council (NDRRMC), Republic of the Philippines. (2011). National Disaster Risk Reduction and Management Plan 2011–2028. Retrieved from www.ndrrmc.gov.ph/attachments/article/41/NDRRM_Plan_2011-2028.pdf

National Disaster Risk Reduction and Management Council (NDRRMC), Republic of the Philippines. (2014) National Disaster Response Plan for Hydro-Meteorological Hazards. Retrieved from https://dromic.dswd.gov.ph/ndrp/

Neria, Y., Nandi, A., & Galea, S. (2008). Post-traumatic stress disorder following disasters: A systematic review. *Psychological Medicine*, *38*, 467–480.

Norris, F., Friedman, M., Watson, P., Byrne, C., Diaz, E., & Kaniasty, K. (2002). 60,000 disaster victims speak: Part I. An empirical review of the empirical literature, 1981–2001. *Psychiatry*, *65*, 207–239.

Panyayong, B., & Pengjuntr, W. (2006). Mental health and psychosocial aspects of disaster preparedness in Thailand. *International Review of Psychiatry*, *18*(6), 607–614.

Philippine Disaster Risk Reduction and Management Act. (2010). Philippine Disaster Risk Reduction and Management Act of 2010: An act strengthening the Philippine disaster risk reduction and management system, providing for the national disaster risk reduction and management framework, and institutionalizing the national disaster risk reduction and management plan, appropriating funds therefor and for other purposes/Philippines Republic Act 10121, S. No 3086 11 No. 6985. Retrieved from http://www.ndrrmc.gov.ph/attachments/article/45/Republic_Act_10121.pdf

Prasetiyawan, Viora, E., Maramis, A., & Keliat, B. A. (2006). Mental health model of care programmes after the tsunami in Aceh, Indonesia. *International Review of Psychiatry*, *18*(6), 559–562.

Rahayu, H., & Ophiyandri. T. (2018). The current context of multi-hazard early warning systems for coastal resilience at the national level: Indonesia. Retrieved from http://cabaret.buildresilience.org/index.php/cabaret-documents

Reifels, L. (2018) Reducing the future risk of trauma: On the integration of global disaster policy within specific health domains and established fields of practice. *International Journal of Environment Resource Public Health*, *15*, 1932.

Roudini, J., Khankeh, H. R., & Witruk, E. (2017, January–June). Disaster mental health preparedness in the community: A systematic review study. *Health Psychology Open*, 1–12. doi:10.1177/2055102917711307

Sagala, S., & Okada, N. (2009). Predictors of intention to prepare for volcanic risks in Mt. Merapi, Indonesia. *Journal of Pacific Rim Psychology*, *3*(2), 47–54.

Satapathy, S. (2012). Mental Health Impacts of Disasters in India: Ex-Ante and Ex-Post Analysis. In Y. Sawada & S. Oum (Eds.), Economic and Welfare Impacts of Disasters in East Asia and Policy Responses. ERIA Research Project Report 2011-8, pp. 425-461. Jakarta: ERIA. Retrieved from https://www.eria.org/Chapter_12.pdf.

Sattler, D. N. (2008). The international tsunami museum: Researchers making a difference in Thailand. *Natural Hazard Observer*, *32*(3), 1–3.

Setiawan, G. P., & Viora, E. (2006). Disaster mental health preparedness plan in Indonesia. *International Review of Psychiatry*, *18*(6), 563–566.

State Gazette of Republic of Indonesia Number 66. (2007). Law of the Republic of Indonesia concerning disaster management/Law no. 24, State Gazette of Republic of Indonesia Number 66. Retrieved from https://www.preventionweb.net/english/policies/v.php?id=10841&cid=80

Thailand Disaster Prevention and Mitigation Act (2007). Thailand Disaster Prevention and Mitigation Act B.E. 2550 [A.D. 2007], Royal Gazette Vol. 124 § Part 52 A (i). Retrieved from http://thailaws.com/law/t_laws/tlaw0397.pdf

Thailand National Disaster Risk Management Plan. (2015). Retrieved from https://www.preventionweb.net/english/policies/v.php?id=54086&cid=170

United Nations Educational, Scientific and Cultural Organization (UNESCO). (2015). *Supporting, Enabling and Empowering Students: Manual on Psychosocial Interventions for Secondary School-aged Students During Disasters and Emergency Situations.* Jakarta, Indonesia: UNESCO Office Jakarta and Japan Official Development Assistance. Retrieved from https://zh.unesco.org/system/files/careers/sees_philippines_supporting_disasters_emergencysituations.pd

United Nations General Assembly (UNGA). (2016). Report of the open-ended intergovernmental expert working group on indicators and terminology relating to disaster risk reduction. Retrieved from https://www.undocs.org/A/71/644

United Nations International Strategy for Disasters Reduction (UNISDR). (2007). Words into action: A guide for implementing the Hyogo framework. Retrieved from https://www.preventionweb.net/publications/view/594

United Nations International Strategy for Disasters Reduction (UNISDR). (2015). Sendai Framework for Disaster Risk Reduction 2015–2030. Retrieved from https://www.preventionweb.net/files/43291_sendaiframeworkfordrren.pdf

Welton-Mitchell, C., James, L. E., Khanal, S. N., & James, A. S. (2018). An integrated approach to mental health and disaster preparedness: A cluster comparison with earthquake affected communities in Nepal. *BMC Psychiatry*, *18*, Article ID 296.

Wolff, P. H., & Fesseha, G. (2005). The Orphans of Eritrea: What are the Choices? *American Journal of Orthopsychiatry*, 75(4), 475–484. http://dx.doi.org/10.1037/0002-9432.75.4.475

World Health Organization (WHO). (2007). The mental health and psychosocial aspects of disaster preparedness (No. SEA-Ment-150): Report of an Intercountry Meeting Khao Lak, Thailand, 20-23 June 2006. New Delhi, India: WHO Regional Office for South-East Asia. Retrieved from https://apps.searo.who.int/PDS_DOCS/B0418.pdf

World Health Organization (WHO). (2011). *Mental health atlas 2011: Indonesia*. World Health Organization. Retrieved from https://www.who.int/mental_health/evidence/atlas/profiles/idn_mh_profile.pdf

World Health Organization (WHO). (2019). *Disaster and mental health*. Dhaka, Bangladesh: World Health Organization Bangladesh. Retrieved from http://www.searo.who.int/bangladesh/disastersandmentalhealth/en/

Yusuf, N. R. (2016). Why Indonesia must invest in mental health. Retrieved from https://www.rappler.com/world/regions/asia-pacific/indonesia/bahasa/englishedition/148788-mental-health-illness

PART III

MENTAL HEALTH AND PSYCHOSOCIAL SUPPORT INTERVENTIONS

CHAPTER 5

PSYCHOLOGICAL FIRST AID: APPLICATION AND ADAPTATION IN SOUTHEAST ASIA

Jason O. Manaois, Chantal Ellis S. Tabo-Corpuz and Andrew G. Heise

ABSTRACT

This chapter reviews the empirical evidence for Psychological First Aid (PFA) in the context and experience of the Southeast Asian (SEA) region. First, this chapter provides the definition and background of PFA and its core principles, to explain the basis for doing PFA as part of an integrated approach to disaster mental health. Second, the existing literature on the effectiveness of PFA is reviewed. Third, this chapter examines the application and adaptation of PFA in SEA. Implications and recommendations are provided at the end of the chapter.

Keywords: Psychological First Aid; Southeast Asia; cultural adaptation; mindfulness; group-based; open-space

According to the United Nations secretariat of the International Strategy for Disaster Reduction (UNISDR, 2010), the frequency of natural hazard disasters has increased over the past years. Southeast Asia is one of the vulnerable regions exposed to natural hazards and threats. Tropical cyclones and earthquakes are a regular occurrence in the region. Climate change has aggravated the intensity and magnitude of these environmental disturbances that lead to rising sea levels,

Resistance, Resilience, and Recovery from Disasters: Perspectives from Southeast Asia
Community, Environment and Disaster Risk Management, Volume 21, 75–93

ISSN: 2040-7262/doi:10.1108/S2040-726220200000021005

urban flooding, landslides and other natural disasters. In addition, there are human-induced hazards caused by political-socioeconomic situations, such as government destabilization and armed conflicts (UNISDR, 2010). These disasters have not only brought destruction and loss of lives, livelihood and property, but have also caused grave suffering in the lives of those who are affected. The mental health of vulnerable populations may be particularly affected by disasters, such as children, people of low socioeconomic status, pregnant or older adults, and people with disabilities. With their limited capacity to function and uncertain access to resources, they may be particularly prone to suffering (UNISDR & UNESCAP, 2010).

Disaster mental health and psychosocial support (MHPSS), which has gained prominence in the past two decades, have opened up new opportunities to explore psychosocial dimensions in relation to well-being (Joseph, 2016). Research has shown that the psychological impact of disasters occurs in all domains of functioning: emotional, cognitive, physical, behavioral, and spiritual (Galea, Nandi, & Vlahov, 2005), though many people are resilient and will eventually return to their pre-disaster level of functioning. On average, 30–40% of disaster victims will experience one or more disorders such as posttraumatic stress disorder (PTSD), depression, or anxiety, with children at greater risk than adults (Galeaet al., 2005). Thus, early intervention is a key component for reducing the risk of developing severe psychological reactions.

Psychosocial support in the field of disaster mental health includes a range of strategies and programs to help individuals and their communities to recover from crises, manage feelings of hopelessness, and develop strengths and skills which can be utilized when facing disaster-related challenges. People are affected in various ways in disaster and emergency situations, requiring various types of support and specific interventions that address their needs. The Inter-agency Standing Committee (IASC, 2007) has developed a layered system of complementary support that meets the needs of different groups, essential in organizing mental health and psychosocial support activities in the community (see Fig. 1). The primary objective is to enable government agencies, community stakeholders, and humanitarian organizations to plan, develop, and coordinate a multi-sectoral response to protect survivors' mental health immediately after emergency situations and improve their psychosocial well-being over time as part of post-disaster recovery intervention. These support interventions include the following phases. *Basic services and security* refers to the notion that responders should address basic survival needs and provide them with a sense of security. It is usually offered to the general population of disaster survivors. *Community and family support* entails providing psychosocial support activities with the aim of promoting, protecting, and aiding the well-being of affected individuals. Many but not all disaster survivors can benefit from this type of intervention. Upon assessment, those individuals who suffer mild psychological distress undergo psychosocial support programs. *Focused, non-specialized supports* entails providing a targeted intervention for individuals or groups that experience a mild to moderate psychological reaction following disaster exposure. *Specialized services* are provided to selected individuals who may suffer from severe psychological disorders. Such individuals

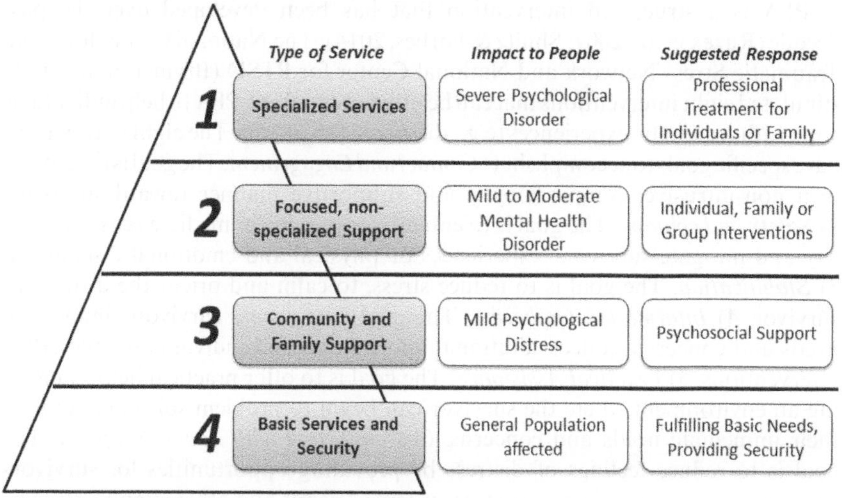

Fig. 1. Inter-agency Standing Committee (IASC) Intervention Pyramid for MHPSS During Emergencies. Adapted from Inter-Agency Standing Committee (IASC) (2007).

may be in need of longer term psychotherapy with a licensed mental health professional, rather than psychosocial support.

It is essential that certain principles are observed in providing MHPSS intervention. IASC (2007) stipulated that human rights and equity must be observed such that fairness in the availability and access to MHPSS support across gender, age, and ethnic groups are considered. Participation must include building local capacities, assisting individuals to be self-sufficient, and strengthening resources already present in the community. An integrated social support system should be highlighted where all stakeholders are engaged (e.g., families, community leaders, church, and local and national government).

WHAT IS PSYCHOLOGICAL FIRST AID?

Psychological first aid (PFA) is a psychosocial support intervention given to communities and families experiencing low to mild distress from disaster. It has become prevalent among first responders for early-stage postdisaster intervention (Shultz & Forbes, 2014). PFA has been defined as a compassionate and supportive presence designed to mitigate acute distress and assess the need for continued mental health care (Everly & Flynn, 2005). As an intervention, it operates within the following core principles: (1) promoting sense of safety, (2) promoting calming, (3) promoting sense of self- and community efficacy, (4) promoting connectedness, and (5) instilling hope (Hobfoll et al., 2007). It has been found that these principles facilitate the process of improving adaptive functioning among people who have experienced disaster-related trauma (Ruzek et al., 2007).

PFA is a structured intervention that has been developed over the past decade (Ruzek et al., 2007; Shultz & Forbes, 2014). The National Center for Child Traumatic Stress Network and National Center for PTSD (Brymer et al., 2006) stipulated eight interventions that can be utilized as part of PFA to help individuals cope with traumatic experiences (e.g., disasters, terrorism). The eight core actions have specific goals to accomplish: 1) *Contact and Engagement*. The goal is to respond in a non-intrusive, compassionate, and supportive manner toward survivors. 2) *Safety and comfort*. The goal is to enhance or provide immediate sense of security and mitigate survivors' experiences of physical and emotional discomfort. 3) *Stabilization*. The goal is to reduce stress, to calm and orient the distraught survivor. 4) *Information Gathering*. The goal is to assess survivors' immediate needs and concerns, collect additional information and individually tailor PFA interventions. 5) *Practical Assistance*. The goal is to offer practical help by creating an environment where the survivor can begin to problem solve and address their immediate needs and concerns. 6) *Connection with Social Supports*. The goal is to reduce feelings of distress by providing opportunities for survivors reconnect with their primary support systems, friends and other support systems in the community. 7) *Coping Information*. The goal is to inform and educate people about stress reactions, coping skills, and resilience in disasters, which can help them cope with the traumatic event. 8) *Linkage with Collaborative Services*. The goal is to link survivors with services needed immediately, while also informing them of services that are available to them in the future, as needed (Forbes et al., 2011; Ruzek et al., 2007; Uhernik & Husson, 2009). From these core actions, other PFA models were also developed and are being used for disaster response. There has been an increasing proliferation of PFA models developed by a range of providers to target a particular group or population (Shultz & Forbes, 2014). Among these are the RAPID-PFA model that was developed by Johns Hopkins University and the World Health Organization's (WHO) PFA – Look, Listen, and Link approach.

JOHNS HOPKINS RAPID–PFA MODEL

The RAPID–PFA model is grounded in evidence-informed and empirically based components designed to address the needs of individuals in acute distress, while at the same time enhancing organizational and community resilience (Everly, Barnett, & Links, 2012). It is composed of the following components: 1) *Rapport and reflective listening* entails connecting with the survivors in a non-intrusive manner by introducing oneself and one's intentions. 2) *Assessment* entails evaluating the basic physical and psychological needs of the survivors. In order to determine their needs, the responder must be able to listen to the story and reaction of the survivors. The interaction would elicit the survivor's current physical and psychological state. It is important to ask specific questions necessary to clarify ambiguous personal reactions of survivors. 3) *Prioritization* is an extension of assessment wherein the responder must be able to identify or focus attention on survivors who are in the greatest need of immediate or emergent care.

The responder is able to assess and identify individuals whose basic functional capacities have been impaired. 4) *Intervention* entails attending to the basic needs of survivors. The responder must be able to address the medical, physical, and psychological needs of survivors. Survivors' experienced psychological distress should be minimized, while their basic level of functionality is restored. 5) *Disposition and follow-up* requires responders to evaluate and make decisions to either end or continue giving intervention. Responders must be able to assist and refer individuals who may need follow-up and/or a higher level of care.

WHO PFA MODEL

The WHO PFA model has three basic action principles: Look, Listen, and Link (WHO, 2011). These action principles help and guide responders on how to safely enter affected communities, approach distressed individuals and understand their needs, and link them with practical support and information. The *Look* action principle involves ensuring that the survivors and responders are in a safe location to conduct PFA. Responders must be able to check for people with obvious urgent basic needs and prioritize individuals with serious distress reactions. It is important to take time assess before offering help. In crisis situations, it is important that responders are calm and have a sense of safety in order to think and respond accordingly. The *Listen* action principle entails approaching and engaging with people who need support. Responders must be able to ask about people's needs and concerns. It is important that responders are able to listen to people and help them to feel calm. The *Link* action principle entails assisting people in addressing their basic needs and access to social services. Responders must be able to help people cope with their problems through providing accurate information and connecting them with their family and other social support. In stressful situations, people may feel vulnerable, isolated or powerless, and unable to access their usual supports. Thus, linking people with practical support is an essential component of PFA. Responders are able to provide support so that survivors are able to help themselves and regain control of their situation. Survivors must be able to identify and re-establish their strengths and coping mechanisms so they are able to effectively address their problems. Save the Children has adapted WHO's PFA, creating a context-specific approach where greater emphasis on understanding children's reactions to crisis and their distinct needs are considered (Dybdal, Melin, & Terlonge, 2013). The training manual for the adaptation emphasizes the role of parents and caregivers and utilizes role play and games as helpful tools in communicating with children (Dybdal et al., 2013).

Table 1 summarizes three PFA models juxtaposed each core components. It appears that all three models are aligned with PFA core principles stipulated by Hobfoll et al. (2007). In addition, they emphasized the need for PFA responders to establish contact and build rapport with disaster survivors. This action signifies the importance of initial PFA engagement to be in a nonintrusive, compassionate and supportive manner. Moreover, the three models share similarities in terms of collecting information and assessing needs to prioritize and tailor-fit

Table 1. Summary of PFA Models.

Model	Core Action (PFA Components)
National Child Traumatic Stress Network and National Center for PTSD	1. Contact and Engagement 2. Safety and Comfort 3. Stabilization 4. Information Gathering 5. Practical Assistance 6. Connection with Social Supports 7. Coping Information 8. Linkage with Collaborative Services
Johns Hopkins RAPID	1. Rapport and Reflective Listening 2. Assessment 3. Prioritization 4. Intervention 5. Disposition and Follow-up
WHO	1. Look 2. Listen 3. Link

PFA intervention at the individual and community-level. The PFA models all highlight the importance of PFA responder's ability to listen actively to survivors needs and to connect survivors to various service providers and social support network. The RAPID model, on the other hand, indicated the need for continued monitoring and follow-up to ensure assistance and referral of survivors in need of higher care. In general, the three PFA models are more similar than different, varying only in implementation steps while keeping within (staying congruent with) the core principles of PFA.

TERMINATING PFA INTERVENTION

PFA can often be a one-time intervention, and responders can only be there to help and assist survivors for a limited time (WHO, 2011). Thus, an important action for PFA responders is deciding when to terminate individualized PFA intervention. Responders must be able to assess whether the person needs continuing help. There is a need to ensure disaster survivors have links with adequate social support and access to other resources needed for their full recovery. Responders should conclude or end their interaction in a positive way ensuring that the person has improved psychological state and functioning. This means that disaster survivors are able to use their own coping skills to recover in the long term.

After doing PFA activities, it is important to formalize the exit from disaster communities. Formalized exit includes: (1) establishing a continuing communication with community leaders, (2) referring vulnerable community members with severe mental health problems (e.g., depression, anxiety, schizophrenia, bipolar) to government social services (e.g., Department of Social Welfare and Development, Department of Health), (3) reporting information (e.g., number of people and area served, PFA intervention used) to local government units and community stakeholders for monitoring purpose and avoid saturation or duplicate intervention, and (4) ensuring disaster communities are linked to agencies doing long-term assistance and rehabilitation. Through this approach, PFA responders show respect and commitment for the full recovery of the disaster community and its members.

EVIDENCE FOR PFA'S EFFECTIVENESS

Although the use of PFA has risen following the September 11, 2001 terrorist attacks in the United States, few studies have investigated these models empirically, perhaps due to methodological complexities (Shultz & Forbes, 2014). Among the empirical articles that do exist, the focus of inquiry has largely been on adaptability of the model for various populations (Brown et al., 2009; Schafer, Snider, & van Ommeren, 2010) and training effects (Lee, You, Choi, Youn, & Shin, 2017), rather than examining whether PFA actually reduces psychological suffering. A handful of recent studies have attempted to close this gap in the literature, providing preliminary support for the effectiveness of PFA.

A pilot study asking college undergraduates to discuss a "personal stress event" found preliminary support for the Johns Hopkins model of PFA (RAPID-PFA) when administered individually (Everly, Lating, Sherman, & Goncher, 2016). Participants were randomly assigned to either the PFA group or a social acknowledgement only group. Participants in the PFA group received a needs assessment, reflective listening, cognitive reframing and other cognitive-behavioral techniques, and normalization of the person's reactions, whereas those in the social acknowledgement group received active listening only. The researchers found a significant reduction of state anxiety and negative mood states in the PFA group, a finding not observed in the social acknowledgement only control group. Importantly, the study also found that anxiety and negative mood states initially increased in both groups while the stress event was discussed, leading the researchers to suggest that the benefit of PFA may have a delayed response. Others have suggested Everly et al. overstated their claims of validation for the model, citing their own statistical analyses of the data that suggest PFA did not show a clear advantage over the control group (Vanhove, Avau, De Buck, Vandekerckhove, 2018). Nevertheless, the study identified a promising intervention warranting further investigation.

Recently, a randomized controlled trial (RCT) compared group-administered RAPID-PFA to a group conversation control condition (Despeaux, Lating, Everly, Sherman, & Kirkhart, 2019). Participants were non-traumatized, undergraduate college students ($n = 119$) who were instructed to watch a distressing video showing violence toward Syrian civilians during the April 22, 2011 attack. Participants were then randomly placed into the PFA condition or the control condition. Those in the PFA condition received reflective listening, normalization, and validation of responses, reframing, and other cognitive-behavioral techniques, and discussion of self-care strategies, in accordance with Johns Hopkins University's standardized RAPID-PFA online course. Participants in the control group were asked to talk among themselves after watching the video, with no specific instructions given. Participants were asked to fill out self-report measures of state-trait anxiety and positive/negative affect at three time points: before the intervention, immediately after the intervention, and 30 minutes post-intervention. The researchers found that group-administered RAPID-PFA was more effective than the control group at reducing state anxiety and acute negative affect immediately after intervention and at 30 minutes post-intervention (Despeaux et al., 2019).

A quasi-experimental pilot study used the school-based PFA, Listen Protect Connect (LPC-PFA) with a sample of middle and high school children who had been potentially exposed to stressful life experiences (Ramirez et al., 2013). Participants had been directly impacted by the 2008 Great Flood of Iowa ($n = 8$) or other potentially traumatizing events such as violence exposure ($n = 12$). PFA was provided by school nurses who had each received three hours of LPC-PFA training. The intervention's three-step approach involved reflective listening and question-asking, assessment and discussion of fears/worries as well as identifying those in need of additional care/support, and reconnecting (family/friends) and reengaging (previously enjoyed activities). The researchers found that LPC-PFA significantly decreased symptoms of depression and posttraumatic stress. Although the validity of the study is limited by not having a control group, 40% of the sample had been exposed to an actual disaster, providing preliminary evidence for the effectiveness of PFA. It has been suggested elsewhere that effectiveness may be further enhanced when those who deliver the intervention include trained, indigenous volunteers who can provide culturally appropriate help (Edwards-Stewart et al., 2011), raising the issue of the importance of culturally adapted PFA.

PFA IN SOUTHEAST ASIA

Research has been conducted about PFA in the Philippines and Brunei. Landoy, Hechanova, Ramos, and Kintanar (2015) conducted a study of the experiences of PFA facilitators in the Philippines. Their study showed that the PFA volunteers' knowledge of PFA, including the goals of PFA, were consistent with core principles of the PFA framework. They described the process as easy, simple, and humane, and had a generally positive experience conducting PFA. Hechanova, Ramos, and Waelde (2015) evaluated the effectiveness of a group-based mindfulness-informed PFA, a Philippine adaptation of PFA. They found that the intervention increased the participants' knowledge of disaster reactions and coping self-efficacy. This finding supports that a culturally adapted PFA still retains its effectiveness when culturally adapted in a non-western context.

Zakaria and Chuemchit (2017) also conducted a study assessing the knowledge, attitude, and practice of PFA of the health professionals working in emergency departments in hospitals in Brunei. They used a questionnaire to measure their health professionals' knowledge, attitude, and practice of PFA. This survey was reviewed by three panel experts on PFA and was pilot tested. They found that 65% of their 123 participants had moderate to high knowledge of PFA. Interestingly, only 11% previously heard of PFA, and yet 51% of the health professionals practiced PFA to a moderate degree. That is, health professionals had moderate knowledge of and practice PFA although they seemed to be unaware that what they are doing was considered PFA. This suggests the need for more training and standardized policies and guidelines, so Southeast Asian (SEA) health professionals can learn more about PFA and its core components.

CULTURAL ADAPTATIONS OF PFA

Culture plays an important role in interventions because it influences people's belief on perceiving what the problem is, or even the existence of the problem itself (Cheung & Snowden, 1990). Culture is, therefore, an important factor to consider when adapting a foreign framework (Berry & Sam, 1997). There are five stages of cultural adaptations of interventions (Barrera Jr, Castro, Strycker, & Toobert, 2013). Stage 1 involves identifying whether the cultural adaptation is essential to the program's outcome, and identifying what elements need to be changed. Stage 2 is where adaptations are made, such as linguistic translations of materials, measurements, and instructions. Stage 3 is pilot testing the adapted program and its measurements. Stage 4 is about refining the study based on the data gathered from stage 3. Stage 5 is the empirical trial of the culturally adapted intervention. Although there is evidence that culturally adapted programs and interventions result in better engagement and retention rates and produce positive outcomes (Baydala et al., 2009; Castro-Olivo, 2014; Kumpfer, Alvarado, Smith, & Bellamy, 2002; Kumpfer, Xie, & O'Driscoll, 2012; Martinez Jr & Eddy, 2005), determining the elements that need to be changed and testing these changes are the challenging processes (Bernal, Jimenez-Chafey, & Domenech Rodriguez, 2009).

Performing psychological services and intervention without considering the cultural background of the client/survivor may lead to harmful results. For example, the Cambodian civil war made Cambodians cautious when receiving help from others. Therefore, establishing trust with the mental health professional should have more importance than the usual rapport-building (Rozée & Van Boemel, 1990; Strober, 1994). To address this concern, van de Put and van der Veer (2005) suggested establishing a good relationship with the *phum* (village) leaders, traditional healers, elderly, and teachers as they are trusted by the local community.

Cultural adaptation of post-disaster intervention has been an important theme in the literature about SEA implementations (Chakrabhand, Panyayong, & Sirivech, 2006; Hechanova, Ramos, et al., 2015; Landoy et al., 2015; Strober, 1994; Sundram et al., 2008; van de Put & van der Veer, 2005). PFA adaptations entail the consideration of cultural factors while retaining the core principles and goals of PFA.

In the Philippine adaptation of PFA, Landoy et al. (2015) listed four main considerations as basis for a Filipino adaptation of PFA. These are the *pagdadala* model, mindfulness, group-based PFA, and open-space activity.

PAGDADALA MODEL

This framework was developed by a Filipino clinical psychologist Edwin Decenteceo (1999). The *pagdadala* translates as burden-bearing for Filipinos. Typically, when disaster strikes, there is chaos and families need to recover all that they have lost. This need becomes a burden, and the survivor bears that burden. For example, after losing a home to a typhoon parents usually find means to provide shelter for their family. In this situation, they are called *nagdadala*, or the ones carrying the burden. The goal of PFA is to raise efficacy so that survivors will be able to recover from the incident. In this manner, PFA teaches Filipinos healthy ways to cope through and eventually lessen the burden. Disaster survivors do not focus sharing their experiences during disaster, although some survivors spontaneously talk about it (Verzosa, 2011), rather the structured conversation featured in pagdadala explores personal strengths, coping strategies, and planning for the future. The *pagdadala* model (Decenteceo, 1999) emphasizes the following major discussion topics for disaster survivors: *dinadala, patutunguhan, pagdadaanan, kalakasan, katulong, and pag-asa. Dinadala* (burden), pertains to the roles, responsibilities, relationships, and experiences that a disaster survivor carries. Survivors elaborate the manner in which they carry their burdens at present; *Patutunguhan* (direction) entails the destination or goals of the burden bearer. Disaster survivors share specific aspirations and future courses of action for alleviating the burden. *Pagdadaanan* (path) pertains to the route taken by the burden bearer. This phase of the conversation focuses on the plan and strategies disaster survivors use in order to bring their role or responsibilities to their desired state or vision. At this point, the conversation is oriented toward making a decision to continue their journey (or moving on), accepting and facing the reality of their situation, and to make concrete solutions to address their predicament. *Kalakasan* (strengths) refers to exploring internal and external resources. After processing the experiences (e.g., stressors, issues, and concerns) of disaster survivors, this phase helps them identify available resources and coping behaviors. *Katulong* (support) involves the ways in which disaster survivors manage their emotions and problems with the support of other people, personal drive, and spirituality. *Pag-asa* (hope), the last phase of the discussion, is intended to provide a sense of hope by focusing on disaster survivors' strengths and their problem-solving strategies. It aims to encourage survivors to focus and work on their goals, and maximize their existing social support. In sum, PFA responders assist disaster survivors to identify and describe their burdens, strategize problem-solving, think of internal–external strengths and coping skills, maximize social support, and have positive expectancies. In essence, the *pagdadala* model and its conversation process is aligned within the PFA principles particularly in the promotion of self-efficacy, utilization of social support, and instillation of hope.

MINDFULNESS

Mindfulness and meditation techniques were included in the adapted version of PFA to help with the stabilization of Filipino survivors (Hechanova, Ramos, et al., 2015). Breath-focused awareness exercises were included in intervention to help the participants become calmer and regulate stress responses. Disaster survivors were taught mindful awareness of their breathing. This was done by instructing survivors to notice physical sensations of breathing. The "letting go exercise" was used to help disaster survivors recognize and relieve experienced anxiety and distress. This exercise involved instructing survivors to visualize (imagine) a balloon or bubble which expands as they inhale and dissolves any sense of tension as they exhale, thus, letting go of any sense of tension felt in the body (Waelde, 2005). The process of doing mindfulness exercises can be repeated PFA between sessions as the need arises. Handouts of additional information and description of mindfulness techniques to help survivors practice both on a daily basis and for short periods of sitting meditation were provided (Hechanova, Ramos, et al., 2015). The use of mindfulness activity is aligned within the PFA principles of promoting calmness and increasing a sense of self-efficacy in handling stressful situations.

GROUP-BASED PFA

There is evidence that survivors in collectivist cultures prefer being in groups than one-on-one sessions because it provides a sense of solidarity. A study in the Philippines suggested that group-based interventions are consistent with the Filipino value of *bayanihan*, which may be roughly translated to cooperative endeavor. This value is reflected in labor mutualization and community members' mobilization during disasters (Galliard, Pangilinan, Cadag, & Le Masson, 2008).

A study in the Philippines describe how group-based PFA is carried out (Hechanova, Ramos, et al., 2015). First, participants should be assessed and profiled for current severity of PTSD symptoms, using a well-validated scale such as the PTSD Checklist for DSM-5 (PCL-5; Weathers et al., 2013) prior to PFA activity. Those participants with significant symptoms above the clinical cutoff score should be grouped together and facilitated by a professional clinician as they may need specialized assistance. However, it is important that group assignments are announced with no reference to PTSD severity to avoid stigma. In addition, clustering the groups based on their profile such as stages of psychosocial development (e.g., children, youth, adults), gender-roles (e.g., mothers, fathers) and vulnerabilities (e.g., persons with disability, sexually abused) might be necessary. To optimize the process of group sharing, an active small group with five to seven members is recommended. Second, in relation to the core PFA action of engagement, PFA small group facilitators invite participants to introduce themselves to the group and share about their current level of functioning. PFA facilitators should be able to establish that anxiety is a normal reaction to stressful situations and traumatic events. Furthermore, facilitators must be able to validate and affirm positive coping strategies used by disaster survivors. This action encapsulates the core principle of promoting self-efficacy. Third, to further deepen the sense of self-efficacy among survivors, psychoeducation about stress and coping is given. PFA facilitators provide information on the stress reaction, coping, and other strategies (e.g., mindfulness, bioenergetic exercises, dietary supplements) that can be practiced by disaster survivors. Group-based PFA is aligned within the PFA principles as it promotes a sense of self- and community efficacy, and connectedness among members.

OPEN-SPACE ACTIVITY

Open-space activities has been used in group-based PFA for survivors who are not yet ready to share their concerns at the time of the intervention (Hechanova, Manaois, & Masuda, 2019; Hechanova, Ramos, et al., 2015). This procedure facilitates engagement for those who are not ready for self-disclosures but who would engage providing support and help to others. The open-space activity consists of three phases: agenda setting, meetings, and plenary sharing (Hechanova et al., 2019). In the first phase, disaster survivors are asked to identify relevant issues and concerns that they wished to discuss. The PFA providers then group the responses into different themes and these become the agenda for the meeting (e.g., medical and hospital concerns, food security, temporary shelters, and financial problems). For each agenda item, a PFA provider is assigned to facilitate each small group. In the meeting phase, disaster survivors are invited to explore and join whatever topic they wished to participate in. They are encouraged to move freely from one meeting space discussion to another. Disaster survivors are asked to identify what available resources they have and to creatively think of possible solutions to their predicament. The PFA providers are then tasked to document what transpired in the meeting space and to write in an easel sheet the suggestions of disaster survivors. This is followed by a plenary sharing of suggested solutions. This phase emphasizes the collective problem-solving approach of finding the best solution as each group's suggestions and ideas are reported to the large group. The group agrees on the most appropriate solution. Disaster survivors are then encouraged to plan to perform activities that foster recovery, such as coping behaviors, problem-solving, and seeking social support. The use of open-space problem-solving is aligned within the PFA principles that include promoting sense of community efficacy, connectedness, and instilling hope among disaster survivors.

CHALLENGES IN IMPLEMENTING PFA IN THE PHILIPPINES

Landoy et al. (2015) described the challenges PFA facilitators faced when conducting PFA in the Philippines. First, there was a lack of protocol for the safe monitoring and follow-up of participants who may have severe psychopathology. Second, there was a lack of conducive venues to conduct PFA because resources were scarce, particularly in a disaster aftermath. Third, there was difficulty in linking the survivors with the appropriate services. Fourth, the country lacked disaster response systems, such that there were no centralized disaster response system set in place. Lastly, only a limited number of individuals were trained in conducting PFA in communities.

These studies suggest the feasibility of adapting PFA in SEA countries while maintaining its effectiveness. Future adaptations should be mindful of cultural values and beliefs that would make PFA more suitable in the local context. Although there are some efforts and initiatives to study and evaluate PFA in SEA countries (see Hechanova, Ramos, et al., 2015; Zakaria & Chuemchit, 2017), there are still areas of improvement in PFA in SEA countries.

APPLICATION OF PFA IN SOUTHEAST ASIA

Research is important in gaining evidence for the effectiveness of PFA in SEA countries. However, lack of research does not equate to non-practice. Some SEA countries practice interventions after disaster strikes and some of the immediate interventions and disaster management plans are quite similar to the elements of PFA. These emergency responses were not labeled as PFA in the literature per se, but they bear some resemblance to the core principles of PFA.

Following natural disasters, government agencies and local Non-government Organizations (NGOs) work together to create a sense of safety and control among the affected community (Chakrabhand et al., 2006; Htay, 2006; Sundram et al., 2008). Temporary shelters are set up, such as in Thailand and Malaysia, where survivors are provided with protection from the weather, clothing, food, other necessities (Chandra, Pandav, & Bhugra, 2006; Krishnaswamy, Subramaniam, Indran, & Low, 2012). There is also an assessment of needs of the survivors (Htay, 2006; Krishnaswamy et al., 2012; Setiawan & Viora, 2006). The needs assessment determines the intervention to be conducted. Coordination is crucial to prevent multiple assessment and to avoid burdening the survivors with too many surveys or interviews. These practices are in line with the core principles of promoting a sense of safety and promoting calming as they address survival needs.

Consequently, the alliance of communities and organizations also creates a more organized response system, making disaster management and planning more efficient in SEA countries. Several countries, including Indonesia, Malaysia, Thailand, and Myanmar, train community leaders to be able to conduct psychosocial interventions (Chakrabhand et al., 2006; Htay, 2006; Krishnaswamy et al., 2012; Setiawan & Viora, 2006). The rationale for focusing on community leaders is that they are the first ones the local community will turn to for support and in times of crisis (Hechanova, Ramos, et al., 2015). Relatedly, the creation of a system of coordination in the local and national setting is helpful in disaster management. This includes collaborations with other agencies and developing committees for the sole function of organizing the different resources (Long, 2001; Setiawan & Viora, 2006). In Singapore, for example, the National Emergency Behavior Management System (NEBMS) was established in 1994 because of the many disasters Singapore faced. NEBMS engage different ministries to provide comprehensive help to the affected areas. These include Ministry of Health (the one primarily in charge of the psychosocial interventions), Singapore Civil Defense Force, Singapore Police Force, Ministry of Community Development and Sports, Ministry of Education, Ministry of Information and Arts, Ministry of Environment, Ministry of Defense, and Prisons Department. The cooperation of the several ministries facilitates inter-agency coordination, making it easier to conduct trainings and exercises across the different ministries (Long, 2001). Having groups and organizations coordinate with one another to create a system of disaster response even before disaster strikes creates a sense of care and solidarity and community efficacy.

In times of disasters, mental health professionals, which typically include psychologists, psychiatrists, doctors, nurses, and social workers, collaborate with one another for a more informed and appropriate plan of action. In many settings, this team approach guides volunteer workers in their duties and conducts trainings on appropriate interventions (Chakrabhand et al., 2006; Htay, 2006; Long, 2001). Their knowledge and expertise facilitate the identification of mental health disorders and needs assessment of families. For example, in Malaysia, disaster responders conducted house visits of disaster survivors, giving the opportunity to allow the families to share their feelings, if the survivors choose to do so. They conducted play therapy for children to identify their mental state. They also trained teachers in schools so they could provide counseling services to their students if needed. After one month, the team and their volunteers conducted another assessment using structured interviews through another round of house visits (Krishnaswamy et al., 2012). In Thailand, a screening questionnaire was given to the survivors. A diagnostic interview was given to those who scored high in the screening tool. Treatment was given to those who were diagnosed with disorders, and community-level psychoeducation is given to those who were not. Survivors who were identified to have severe symptoms even after treatment were referred to the proper facilities. Beyond providing the proper mental health care of the survivors, this practice relates to the PFA principle of regaining their sense of self and instilling hope.

Disaster response practices of SEA countries have a strong resemblance to PFA. Though literature has been limited in terms of PFA studies, the goals and principles of PFA are represented through the practice of providing immediate needs (e.g., temporary shelters and relief goods), identifying short-term and long-term needs through assessment, cooperation among the different organizations to have a more cohesive response to the different needs, providing trainings for service providers and psychoeducation for affected individuals and families, and referral to the appropriate organizations and clinics. More research should be done for PFA in SEA countries, but it seems evident that PFA can be adapted to SEA countries.

IMPLICATIONS AND RECOMMENDATIONS

Strengthen PFA Evaluation

There is a growing need to provide evidence of PFA effectiveness done in SEA. Varied adopted PFA programs have been observed in the region. These programs partially adhere to the core principles of PFA stipulated by the National Center for Child Traumatic Stress Network and National Center for PTSD (Brymer et al., 2006). However, because the programs are not studied and evaluated as PFA, they do not strictly follow the suggested action points of PFA studies. This lack of consistent programing can present challenges for systematic program evaluation.

Conduct Research About Local Adaptations of PFA

Mental health professionals should conduct research that evaluates locally adapted PFA. Such studies would make their programs more evidence-based to further enhance not only PFA literature but also similar psychosocial interventions in their countries. Future research may explore the long-term effects of PFA intervention for addressing psychosocial needs of disaster survivors. Moreover, this type of research helps raise global awareness of the influences of SEA culture on mental health and psychosocial intervention programs.

Although a review suggests that there is general evidence for the effectiveness of PFA (Fox et al., 2012), studies conducted in SEA have not directly measured the core principle of PFA, though some studies have compared pre- and post-tests score difference of participants' organizational support, coping skills, self-efficacy, resilience, and posttraumatic stress symptoms (Hechanova et al., 2019; Hechanova, Ramos, et al., 2015). Thus, it is recommended to construct and validate PFA outcome assessments that measure its core principles of (1) promoting sense of safety, (2) promoting calming, (3) promoting sense of self- and community efficacy, (4) promoting connectedness, and (e) instilling hope. On the other hand, there are studies that investigated the level of knowledge, attitude, and helping behavior practices among professionals who conduct PFA (Zakaria & Chuemchit, 2017; Kitchener & Jorm, 2002). However, these studies did not measure actual mental health impact of conducting PFA with disaster survivors. Therefore, there is a need to develop a common tool for monitoring and evaluating PFA activity for research and documentation clarity. Moreover, RCT research can be done to establish PFA effectiveness relative to other types of intervention.

Increase Capability-building Initiatives

Enhancing the readiness of communities and institutions to respond to the postdisaster psychosocial needs of its members is crucial. Relevant activities would include increasing the knowledge, skills, and confidence of community PFA providers by providing training programs and materials prior to a disaster. Training programs should include PFA training, so that when the need arises, local community providers are already equipped with the necessary skills and knowledge to implement PFA. In addition, coordination among organizations or communities prior to disasters will make their system of postdisaster response more cohesive and efficient.

Initiate Knowledge Sharing Among Communities

Communicating best practices of PFA engagements in communities should be strengthened. Sharing expertise and resources can be initiated to foster collaboration and intensify PFA programs. Feedback from various communities may also enable local PFA trainers to improve their adapted PFA program.

Institutionalize PFA Activities

Government agency (e.g., Department of Health) must be able to monitor PFA activities in the community to protect people from health intervention malpractice. Having one authority in the implementation and training of PFA makes it easier for the country to take charge of the different NGOs and local government units. A system of reporting and documentation can be established to formalize PFA as a legitimate early disaster intervention.

Promote and Strengthen Resilience Interventions

Beyond conducting PFA for immediate disaster response, resilience interventions are needed to help communities adapt better when natural disasters strike, especially given the vulnerable geographic locations of many SEA countries. An example would be the *Katatagan* (Resiliency) intervention that was developed and administered in the Philippines to survivors of Typhoon Haiyan (Hechanova, Ramos, et al., 2015). Other psychosocial interventions can include psychoeducation concerning the physical and emotional symptoms of postdisaster reactions, and learning healthy coping skills and adaptive behaviors. In this way, not only are the immediate needs addressed through PFA, long-term life skills can be learned by the community for better mental and emotional response to future challenges, be it brought by disasters or otherwise.

CONCLUSIONS

In summary, given the dearth of empirical studies on its effectiveness, there is a need to provide further evidence of the efficacy of PFA. Cultural adaptation of PFA for SEA needs to be further validated and systematized. Most research reviewed in this chapter has focused on the design of PFA interventions. However, the region lacks mental health providers and resources to service its huge population. Thus, there is an urgency to create a pool of PFA facilitators as first responders in times of disasters. There is a need to establish standards of competencies of trained PFA providers to ensure quality of service and to protect disaster survivors from malpractice.

REFERENCES

Barrera, M., Jr., Castro, F. G., Strycker, L. A., & Toobert, D. J. (2013). Cultural adaptations of behavioral health interventions: A progress report. *Journal of Consulting and Clinical Psychology*, *81*(2), 196.

Baydala, L. T., Sewlal, B., Rasmussen, C., Alexis, K., Fletcher, F., Letendre, L., … Kootenay, B. (2009). A culturally adapted drug and alcohol abuse prevention program for Aboriginal children and youth. *Progress in Community Health Partnerships: Research, Education, and Action*, *3*(1), 37–46.

Bernal, G., Jiménez-Chafey, M. I., & Domenech Rodríguez, M. M. (2009). Cultural adaptation of treatments: A resource for considering culture in evidence-based practice. *Professional Psychology: Research and Practice*, *40*(4), 361.

Berry, J. W. & Sam, D. L. (1997). Acculturation and adaptation. *Handbook of Cross-Cultural Psychology*, *3*(2), 291–326.

Brown, L. M., Bruce, M. L., Hyer, K., Mills, W. L., Vongxaiburana, E., & Polivka-West, L. (2009). A pilot study evaluating the feasibility of psychological first aid for nursing home residents. *Clinical Gerontologist*, *32*, 293–308. https://doi.org/10.1080/07317110902895317

Brymer, M., Jacobs, A., Layne, C., Pynoos, R., Ruzek, J., Steinberg, A., … Watson, P. (2006). *Psychological first aid: Field operations guide* (2nd ed.). National Child Traumatic Stress Network and National Center for PTSD. Retrieved from www.nctsn.org and www.ncptsd.va.gov.

Castro-Olivo, S. M. (2014). Promoting social-emotional learning in adolescent Latino ELLs: A study of the culturally adapted Strong Teens program. *School Psychology Quarterly*, *29*(4), 567.

Chandra, V., Pandav, R., & Bhugra, D. (2006). Mental health and psychosocial support after the tsunami: Observations across affected nations. *International Review of Psychiatry*, *18*(3), 205–211.

Chakrabhand, S., Panyayong, B., & Sirivech, P. (2006). Mental health and psychosocial support after the tsunami in Thailand. *International Review of Psychiatry*, *18*(6), 599–605.

Cheung, F. K., & Snowden, L. R. (1990). Community mental health and ethnic minority populations. *Community Mental Health Journal*, *26*(3), 277–291.

Decenteceo, E. T. (1999). The Pagdadala model in counseling and therapy. *Philippine Journal of Psychology*, *32*(2), 89–104.

Despeaux, K. E., Lating, J. M., Everly, G. S., Sherman, M. F., & Kirkhart, M. W. (2019). A randomized controlled trial assessing the efficacy of group psychological first aid. *The Journal of Nervous and Mental Disease*, *207*, 626–632. https://doi.org/10.1097/NMD.0000000000001029

Dybdal, A., Melin, M., & Terlonge, P. (2013). *Save the children psychological first aid training manual for child practitioners*. Denmark: Save the Children.

Edwards-Stewart, A., Ahmad, Z. S., Thoburn, J. W., Furman, R., Lambert, A. J., Lauren, S., & Gunn, G. (2011). Introducing embedded indigenous psychological support teams: A suggested addition to psychological first aid in an international context. *International Journal of Emergency Mental Health*, *14*, 289–296.

Everly, G., Barnett, D. J., & Links, J. M. (2012). The Johns Hopkins model of psychological first aid (RAPID - PFA): Curriculum development and content validation. *International Journal of Emergency Mental Health*, *14*(2), 95-103.

Everly, G. S., Jr., & Flynn, B. W. (2005). Principles and practice of acute psychological first aid after disasters. In G. S. Everly Jr. & C. L. Parker (Eds.), *Mental health aspects of disasters: Public health preparedness and response, revised* (pp. 79–89). Baltimore, MD: Johns Hopkins Center for Public Health Preparedness.

Everly, G. S., Lating, J. M., Sherman, M. F., & Goncher, I. (2016). The potential efficacy of psychological first aid on self-reported anxiety and mood: A pilot study. *The Journal of Nervous and Mental Disease*, *204*, 233–235. https://doi.org/10.1097/NMD.0000000000000429

Forbes, D., Lewis, V., Varker, T., Phelps, A., O'Donnell, M., Wade, D. J., … Creamer, M. (2011). Psychological first aid following trauma: Implementation and evaluation framework for high-risk organizations. *Psychiatry*, *74*, 224–239. doi:10.1521/psyc.2011.74.3.224

Fox, J. H., Burkle, F. M., Bass, J., Pia, F. A., Epstein, J. L., & Markenson, D. (2012). The effectiveness of Psychological First Aid as a disaster intervention tool: Research analysis of peer-reviewed literature from 1990-2010. *Disaster Medicine and Public Health Preparedness*, *6*(3), 247–252. doi:10.1001/dmp.2012.39

Galea, S., Nandi, A., & Vlahov, D. (2005). The epidemiology of post-traumatic stress disorder after disasters. *Epidemiologic Reviews*, *27*, 78–91.

Galliard, J. G., Pangilinan, M. R., Cadag, J. R., & Le Masson, V. (2008). Living with increasing floods: Insights from a rural Philippine community. *Disaster Prevention and Management*, *17*(3), 383–395.

Hechanova, M. R. M., Manaois, J. O., & Masuda, H. V. (2019). Evaluation of an organization-based psychological first aid intervention. *Disaster Prevention and Management*, *28*(3), 401–411. https://doi.org/10.1108/DPM-10-2018-0330

Hechanova, R., Ramos, P., & Waelde, L. (2015). Group-based mindfulness informed psychological first aid after Typhoon Haiyan. *Disaster Prevention and Management*, *24*(5), 610–618. https://doi.org/10.1108/DPM-01-2015-0015

Hechanova, M. R. M., Waelde, L. C., Docena, P. S., Alampay, L. P., Alianan, A. S., Flores, M. J. B., & Melgar, M. I. E. (2015). The development and initial evaluation of Katatagan: A resilience intervention for Filipino disaster survivors. *Philippine Journal of Psychology*, *48*(2), 105–131.

Hobfoll, S. E., Watson, P. E., Ruzek, J. I., Bryant, R. A., Brymer, M. J., Pynoos, R. S., ... Ursano, R. J. (2007). Five essential elements of immediate and mid-term mass trauma intervention: Empirical evidence. *Psychiatry*, *70*(4), 283–315. doi:10.1521/psyc.2007.70.4.283

Htay, H. (2006). Mental health and psychosocial aspects of disaster preparedness in Myanmar. *International Review of Psychiatry*, *18*(6), 579–585.

Inter-agency Standing Committee (IASC). (2007). *IASC Guidelines on Mental Health and Psychosocial Support in Emergency Settings*. Geneva: IASC. Retrieved from https://interagencystandingcommittee.org/system/files/iasc_guidelines_on_mental_health_and_psychosocial_support_in_emergency_settings.pdf

Joseph, J. (2016). Disaster mental health vs psychosocial well being: The need for a true paradigm change. *Austin Journal of Emergency and Critical Care Medicine*, *3*(2), 1048.

Kitchener, B. A., & Jorm, A. F. (2002). Mental health first aid training for the public: Evaluation of effects on knowledge, attitudes and helping behavior. *BMC Psychiatry*, *2*(1), 10. https://doi.org/10.1186/1471-244X-2-10

Krishnaswamy, S., Subramaniam, K., Indran, T., & Low, W. Y. (2012). The 2004 tsunami in Penang, Malaysia: Early mental health intervention. *Asia Pacific Journal of Public Health*, *24*(4), 710–718.

Kumpfer, K. L., Alvarado, R., Smith, P., & Bellamy, N. (2002). Cultural sensitivity and adaptation in family-based prevention interventions. *Prevention Science*, *3*(3), 241–246.

Kumpfer, K. L., Xie, J., & O'Driscoll, R. (2012). Effectiveness of a culturally adapted Strengthening Families Program 12–16 years for high-risk Irish families. *Child and Youth Care Forum*, *41*(2), 173–195.

Landoy, B.V., Hechanova, M.R., Ramos, P.A., & Kintanar, N.S. (2015). The application and adaptation of psychological first aid: The Filipino psychologists' experience after typhoon Haiyan. *Philippine Journal of Psychology*, *48*(2), 81–104.

Lee, J.-S., You, S., Choi, Y.-K., Youn, H., & Shin, H. S. (2017). A preliminary evaluation of the training effects of a didactic and simulation-based psychological first aid program in students and school counselors in South Korea. *PLoS One*, *12*(7), e0181271. https://doi.org/10.1371/journal.pone.0181271

Long, F. Y. (2001). Psychological support in civil emergencies: The national emergency behavior management system of Singapore. *International Review of Psychiatry*, *13*(3), 209–214.

Martinez Jr, C. R., & Eddy, J. M. (2005). Effects of culturally adapted parent management training on Latino youth behavioral health outcomes. *Journal of Consulting and Clinical Psychology*, *73*(5), 841.

Ramirez, M., Harland, K., Frederick, M., Shepherd, R., Wong, M., & Cavanaugh, J. E. (2013). Listen protect connect for traumatized schoolchildren: A pilot study of psychological first aid. *BMC Psychology*, *1*, 1–9.

Rozée, P. D. & Van Boemel, G. (1990). The psychological effects of war trauma and abuse on older Cambodian refugee women. *Women & Therapy*, *8*(4), 2350.

Ruzek, J. I., Brymer, M. J., Jacobs, A. K., Layne, C. M., Vernberg, E. M., & Watson, P. J. (2007). Psychological first aid. *Journal of Mental Health Counseling*, *29*, 17–49. doi:10.1037/a0012663

Schafer, A., Snider, L., & van Ommeren, M. (2010). Psychological first aid pilot: Haiti emergency response. *Intervention*, *8*, 245–254. https://doi.org/10.1097/WTF.0b013e32834134cb

Setiawan, G. P. & Viora, E. (2006). Disaster mental health preparedness plan in Indonesia. *International Review of Psychiatry*, *18*(6), 563–566.

Shultz, J. M., & Forbes, D. (2014). Psychological first aid rapid proliferation and the search for evidence. *Disaster Health*, *2*(1), 3–12. doi:10.4161/dish.26006

Strober, S. B. (1994). Social work interventions to alleviate Cambodian refugee psychological distress. *International Social Work*, *37*(1), 23–35.

Sundram, S., Karim, M. E., Ladrido-Ignacio, L., Maramis, A., Mufti, K. A., Nagaraja, D ... Wahab, M.A. (2008). Psychosocial responses to disaster: An Asian perspective. *Asian Journal of Psychiatry*, *1*(1), 7–14.

Uhernik, J. A. & Husson, M. A. (2009). Psychological first aid: An evidence informed approach for acute disaster behavioral health response. *Compelling Counseling Interventions: VISTAS*, 200(9), 271–280.

United Nations secretariat of the International Strategy for Disaster Reduction (UNISDR) (2010). Synthesis Report on Ten ASEAN Countries Disaster Risks Assessment. Retrieved from https://www.unisdr.org/files/18872_asean.pdf

United Nations secretariat of the International Strategy for Disaster Reduction (UNISDR) and Economic & Social Commission for Asia and the Pacific (UNESCAP). (2010). Protecting Developing Gains, Reducing Disaster Vulnerability and Building Resilience in Asia and the Pacific: The Asia Pacific Disaster Report 2010. Retrieved from https://www.unisdr.org/we/inform/publications/16132

van de Put, W. & van der Veer, G. (2005). Counselling in Cambodia: Cultural competence and contextual costs. *Intervention*, *3*(2), 87–96.

Vanhove, A.C., Avau, B., De Buck, E., & Vandekerckhove, P. (2018). Commentary on Everly et al. (2016). *The Journal of Nervous and Mental Disease*, *206*, 390. https://doi.org/10.1097/NMD.0000000000000812

Verzosa, L. (2011). Disasters: The Philippine experience. In. L. Ignacio (Ed.), *Ginhawa: Well-being in the aftermath of disasters* (pp. 3–47). Quezon City: Flipside Digital Content Company, Inc.

Waelde, L. C. (2005). *Inner resources for stress. Treatment manual and materials*. Palo Alto, CA: Inner Resources Center, Palo Alto University.

Weathers, F. W., Litz, B. T., Keane, T. M., Palmieri, P. A., Marx, B. P., & Schnurr, P. P. (2013). The PTSD Checklist for DSM-5 (PCL-5) – Standard [Measurement instrument]. Available from https://www.ptsd.va.gov/

World Health Organization. (2011). Psychological first aid: guide for field workers. World Health Organization.

Zakaria, K. & Chuemchit, M. (2017). Knowledge, attitude and practice of psychological first aid among health professionals working in accident & emergency department in Brunei Darussalam. *Journal of Health Research*, *31*(Suppl. 1), s99–s105.

CHAPTER 6

MINDFULNESS INTERVENTIONS FOR DISASTER RESILIENCE IN SOUTHEAST ASIA

Adriana Panting, Andrew G. Heise,
Ma. Regina M. Hechanova and Lynn C. Waelde

ABSTRACT

This chapter summarizes the literature on mindfulness and its impact of postdisaster response. Although the use of mindfulness is still in its infancy in Southeast Asia (SEA), initial studies suggest it has potential as a means to address survivors' posttrauma symptoms. Given cultural nuances such as a discomfort in emotional expression and shame, mindfulness is non-intrusive and encourages non-judgmental acceptance. Mindfulness has been used in group settings which is congruent with the region's collectivist orientation. In addition, given the importance of spirituality, we suggest that mindfulness may be an inclusive approach that is familiar and acceptable to SEA survivors.

Keywords: Mindfulness; meditation; Southeast Asia; trauma; disaster; training

Southeast Asia (SEA) encompasses a geographic area south of China and east of the subcontinent of India and includes 11 countries: Cambodia, Thailand, Singapore, Laos, Vietnam, Burma (Myanmar), Indonesia, Malaysia, Brunei, Timor Leste, and the Philippines (Center for Southeast Asian Studies, 2019).

Resistance, Resilience, and Recovery from Disasters: Perspectives from Southeast Asia
Community, Environment and Disaster Risk Management, Volume 21, 95–108

ISSN: 2040-7262/doi:10.1108/S2040-726220200000021006

The World Health Organization's list of SEA countries includes Bangladesh, Bhutan, South Korea, India, Maldives, Nepal, and Sri Lanka (World Health Organization Regional Office for South-East Asia, 2019). The millions of people living in the region are prone geographically to natural hazards of several types including typhoons, earthquakes, tsunamis, flooding, volcanoes, forest fires, and landslides (UNISDR & World Bank, 2010). Several SEA regional studies have examined the positive effects of mindfulness on anxiety disorders (Kim et al., 2009), women's mental health (Lee & Bang, 2010), insomnia (Yook et al., 2008), stress management (Phang et al., 2015) and other ailments, but few studies have applied mindfulness interventions in the context of natural hazards. Mindfulness interventions have been found in recent years to reduce trauma-related symptoms, but only a handful of studies have examined its use for disaster survivors. The purpose of this chapter is to establish a rationale for the use of mindfulness interventions for the treatment of stress- and trauma-related symptoms common in those who experience disasters and to review the small body of literature utilizing these techniques in SEA.

POSTDISASTER STRESS RESPONSES

Those that are exposed to disasters typically experience some level of distress. For the majority of survivors, increased symptoms or impairments related to stress occur but do not reach the level of a diagnosable mental health issue. The effects of disasters often include impaired mental health of the individuals that suffered through it. For a substantial minority of those affected, the mental health impairments reach the level of being able to be diagnosed as a mental health issue. For others, pre-existing mental health conditions can be exacerbated by the stress of a disaster (North & Pfefferbaum, 2013).

STRESS RESPONSES

The psychological effects of disaster are often hard to address because unlike physical injuries, they are often not apparent and can take time to develop further and become more visible. Individuals that live through disasters often experience increased emotional distress, psychosocial issues, worry, pessimism, sadness, sleep issues, bereavement and complicated grief reactions, impaired daily functioning, negative attitudes and beliefs regarding safety and the future (North & Pfefferbaum, 2013), and a reduced quality of life (Descilo et al., 2010). A systematic review of 569 unique research articles regarding provision of stress reduction and mental health interventions showed that 40% of individuals in crisis that presented for assessment at relief aid centers and shelters had pre-existing mental health disorders and experienced an increase in symptoms since the disaster occurred (North & Pfefferbaum, 2013). It is notable that the type of disaster that occurred can also impact the level of impairment or the types of impairments

(Briere & Elliott, 2000). For example, those who lived through a natural hazard but were not exposed to death or injury were less likely to experience fear of death or injury. This implies that some psychological changes or symptoms might be influenced by the disaster type.

PROGRESSION TO PSYCHOPATHOLOGICAL DISORDERS

It was recently estimated that consequences of natural hazards in Asia included posttraumatic stress disorder (PTSD), occurring in 8.7–57.3% of survivors (Udomratn, 2008). The systematic review of disaster mental health studies conducted by North and Pfefferbaum (2013) showed that a significant minority of those who presented for assessment at relief aid centers and shelters also had clinical disorders related to poor adjustment and high levels of stress. A study of refugee survivors from coastal fishing villages in India affected by the 2004 Indian Ocean Tsunami eight months after the disaster found that the survivors experienced significant postdisaster impairments as it related to symptoms of, depression, and anxiety. The survivors commonly endorsed experiencing repeated mental images of faces drowning in the flood, insomnia, over-reactivity to the sound of the ocean, difficulty experiencing positive feelings during day-to-day activities, and feeling they were re-experiencing the disaster. They also reported feeling constant, intense, and unbearable sadness, pessimism about the future, anxiety, terror, panic, and worry (Descilo et al., 2010).

Additional psychiatric issues commonly seen in disaster survivors also included major depression and substance use disorders. In children, risk factors for the development of more intense trauma symptoms included being separated from a parent or caregiver and living in emergency relief centers, shelters, or trailers rather than living in their own home. Separation from a caregiver has been proven to be especially traumatic for children during disasters because parental support is crucial for providing emotional relief and a sense of normalcy and security (Osofsky, Osofsky, Kronenberg, Brennan, & Hansel, 2009). An investigation of Chinese survivors following the Wenchuan earthquake of 2008 found that bereavement after natural hazards increases the risk for the development of clinical depression in adults and children (Tang, Liu, Liu, Xue, & Zhang, 2014). These findings imply that those with lower access to education and resources may further be marginalized and experience more serious health issues following disaster.

It is notable that there are disparities among those who develop mental health conditions following disasters; those with higher education levels are often found to use more adaptive coping methods due to privileges involving more access to social resources. Higher education was associated with less depression following disaster, as seen in a study of Chinese survivors following the Wenchuan earthquake of 2008 (Tang, Liu, Liu, Xue, & Zhang, 2014).

SECONDARY CONSEQUENCES OF POSTDISASTER STRESS

The challenges that survivors face after disasters can be harder to overcome due to the impaired functioning that disaster can bring. Many individuals that live through disasters are faced with the task of rebuilding their lives after subsequentially experiencing loss of property, lives, community, family, and economic systems as well as great emotional loss (Piyasil et al., 2008). The impact of stress associated with experiencing disasters can lead to problems returning to the life they once knew, often involving a lifestyle they need to continue to survive. Refugee survivors from coastal fishing villages in India affected by the 2004 Indian Ocean Tsunami reported being triggered emotionally and experiencing intense physical and emotional reactions to the sound of the ocean. Hearing the ocean would trigger in them feelings of anxiety, terror, and panic and would trigger re-experiencing of the disaster. This prevented many survivors from returning to their jobs as fisherman, causing further devastation and stress to individuals and families affected by the tsunami (Descilo et al., 2010). Adults who struggle to find employment after a disaster are more likely to develop symptoms related to depressive disorders. This mental health burden further marginalizes survivors, often leaving them unable to support themselves and their families in the way they could before the disaster (Tang et al., 2014).

As Osofsky et al. (2009) reviewed, many affected by disasters are forced to adapt to life in a new environment, often being one created for aid or relief. The impact of stress associated with experiencing disasters can lead to problems adapting to new living arrangements they seek refuge in. These new conditions often include living in small, overcrowded trailers, centers, or communities with limited space, resources, and autonomy to start and maintain daily routines such as cooking. These transitional living communities that become permanent or longer lasting than intended also have limited privacy, communal, and recreational space, including space for children to engage in activities that could provide stress relief such as play. These living conditions and heightened stress responses often lead to conflicts between family members or others residing in the space. Not being able to achieve a sense of normalcy prolongs and increases the sense of loss for children, adults, and families affected by the disaster, preventing stability that can be provided by regular routines.

The health impairments associated with disaster can progress or manifest differently over time. After a disaster the medical needs of survivors are often neglected or inadequately attended to which leads to higher rates of health complications and indirect morbidity. Those affected by disaster often endorse lower health related quality of life (Jia et al., 2010). Many who have survived disasters experience physical manifestations of stress such as somatic pain or stress-related tension that can impair their ability to engage in occupational or social functioning. Many disasters survivors use substances to cope with the impact of stress and disruption of daily life and routine associated with disasters. This was seen in individuals across the lifespan (children to adults) in a study conducted in with survivors of the 2004 tsunami in Thailand (Piyasil et al., 2008). Addiction can prevent adaptive occupational and social functioning and can lead to increased health issues and impairments.

The stressors and challenges that come with rebuilding life and healing after disaster can prove to be unbearable for many. A comparison of suicide rates 30 years prior to and 10 years following the Taiwan earthquake in 1999 showed that adults 45 years or older had a higher suicide rate when compared to both adults who were not affected by the disaster and the younger adults who were exposed to it (Chen et al., 2016). A suicide can further traumatize a family or loved ones who have experienced disaster, furthering the negative effects of disaster exposure. These findings demonstrate additional reasons for the dire need for effective interventions to reduce the impairments experienced after disasters.

PREVENTING PROGRESSION OF POSTDISASTER EFFECTS

The psychological effects of disaster are often hard to address because unlike the physical injuries and devastation, they are often not apparent and can take time to progress. The effects of some disasters on psychosocial functioning have also been shown to last for long periods of time when untreated, sometimes decades or longer (Briere & Elliott, 2000). Rather than wait for the onset of mental health disorders or worsening of stress symptoms, effective, early interventions can help those affected adopt perspectives and behaviors that may reduce the development of ensuing challenges and disorders (Dyregrov & Regel, 2012). Early interventions can involve efforts for increasing coping and stress reduction after the occurrence of traumatic events that pose high likelihood of experiencing psychiatric distress, symptoms of PTSD, or grief-related issues for a substantial number of those affected, like a disaster.

Traumatic experiences may result in negative thinking and rumination over negative content with little awareness of this tendency (Myers & Wells, 2015). This engagement in rumination leads to increased stress, anxiety, and depression (Aldao, Nolen-Hoeksema, & Schweizer, 2010; Guendelman, Medeiros, & Rampes, 2017). Rumination may also result in skewed conclusions about the information being processed; many individuals who repeatedly process a trauma will often develop the belief that they are completely powerless (Myers & Wells, 2015). The type or severity of disaster does not affect how long symptoms (no matter how mild or severe) persist after the disaster (Briere & Elliott, 2000), but effective, early interventions might. Early intervention can reduce the tendency to negatively or inaccurately judge one's own reactions to the event, sense of power, ability to cope, and resourcefulness. Intervention efforts can meaningfully counteract the development and maintenance of posttraumatic stress reactions and can promote the adoption of perspectives that serve to empower affected people (Dyregrov & Regel, 2012).

Individuals with higher distress tolerance are less likely to perceive and process symptoms of stress related to exposure to traumatic events (Vujanovic, Bonn-Miller, Potter, Marshall, & Zvolensky, 2011). When faced by traumatic experiences, individuals with higher ability to tolerate distress are less likely to perseverate on cues of stimuli that would increase symptoms of stress or they will interpret the stress as not surpassing their coping ability (Vujanovic et al., 2011). Those with a low distress tolerance are more likely to use maladaptive and harmful coping styles to avoid distressing stimuli and feelings that accompany it (Vujanovic et al., 2011). A study conducted with survivors of the 2004 tsunami in Thailand found that those with higher distress tolerance or ability to accept new and life-changing situations showed less emotional and physical signs consequences of stress related to the traumatic event (Piyasil et al., 2008). Early intervention to increase distress tolerance, reduce rumination, and improve physiological stress regulation may improve outcome among disaster survivors; mindfulness interventions may provide these benefits in a way that can be culturally adapted and effective.

MINDFULNESS

The use of mindfulness has origins that can be seen historically in practices related to spirituality and religion. Currently, its secular use has become more popular, highlighting self-regulation of attention as a normative neurodevelopmental process rather than an emphasis on spiritual or religious concepts (Waelde & Thompson, 2016). When conceptualized secularly, mindfulness involves an ongoing process of redirecting one's attentional focus to experiences occurring in the present moment while maintaining an outlook of non-judgment and curiosity. As a result of present moment focus that occurs through mindfulness, an increase of awareness of mental processes, such as changing thoughts, emotions, and sensations can be observed (Bishop et al., 2004).

Mindfulness is thought of as an innate cognitive process that can be further developed and enhanced with regular practice (Bishop et al., 2004; Waelde & Thompson, 2016). Individuals can engage in mindfulness practices in a formal way, setting aside time to practice and learn the skills involved, but mindfulness also refers to a trait like ability to maintain present moment awareness in a non-reactive, non-judgmental way. Both of these occurrences are supported through neurological research data on bottom-up and top-down processing (Guendelman et al., 2017). Mindfulness is associated with the ability to be more focused in the present moment, rather than ruminating over events of the past or worrying about possibilities of the future.

Formal mindfulness practice entails purposefully finding a time and space to practice, often with an intention to reduce stress or increase relaxation (Chiesa & Serretti, 2009). Formal practice is commonly thought of to occur during a longer seated meditation practice, and practice in daily life is often used in more brief practices that can occur in various settings during a wide range of activities (Waelde et al., 2019). These settings include at home (Brown & Ryan, 2003), in classrooms (Bellinger, DeCaro, & Ralston, 2015), workplaces (Aikens et al., 2014), postdisaster settings (Waelde, Hechanova, Ramos, Macia, & Moschetto, 2017), and jails and prisons (Auty, Cope, & Liebling, 2017).

The activities that can be completed while practicing mindfulness also vary; mindfulness has been incorporated into tasks of daily living such as brushing teeth, cleaning (Bögels & Restifo, 2014), walking (Teut et al., 2013), coloring (Thompson & Waltz, 2010), playing sports (Netz & Lidor, 2003), and more. In fact, the item, process, or task that is the focus of attention during mindfulness practice can differ depending on preferences or habits of the person engaging in the practice. The portability and variability in use allows for the practice of mindfulness to be generalizable to different individuals with varied backgrounds, interests, and daily habits (Waelde et al., 2019).

RATIONALE FOR MINDFULNESS FOR POSTDISASTER STRESS RESPONSES

Mindfulness can be a protective factor that can reduce common effects of traumatic or challenging experiences and stress (Weinstein, Brown, & Ryan, 2009). Individuals affected by trauma learn that rather than being overwhelmed by trauma reactions when they arise, they can let them go (Waelde, 2015). Mindfulness promotes the ability to experience feelings of distress without extensive engagement and rumination (Grecucci, Pappaianni, Siugzdaite, Theuninck, & Job, 2015). Although individuals affected by traumatic stress may always have some reaction to trauma-related cues, the goal of mindfulness and related practices is to promote distress tolerance and better self-regulate physiological stress reactions (Waelde, 2015; Waelde, Meyer, Thompson, Thompson, & Gallagher-Thompson, 2017). As individuals increasingly engage in mindfulness, their levels of anxiety and stress decrease (Kiken & Shook, 2012; Mackenzie, Poulin, & Seidman-Carlson, 2006; Mahoney, Segal, & Coolidge, 2015). The ability to tolerate and let go of stress reactions can be a valuable tool for those affected by disaster and ensuing stress responses, whether or not it reaches the level of a diagnosable disorder.

Mindfulness practice encourages individuals with low distress tolerance to develop a non-reactive acceptance of difficult or stressful emotional experiences. As this occurs, one's perception of stress can change. Accepting sensations of distress (instead of avoiding or suppressing them) can allow one to processes the unpleasant stimuli or internal sensations. Doing so, allows one to become fully aware of the normal, transitory, and manageable nature of stress (Mahoney et al., 2015), as their one's ability to respond adaptively to distress is strengthened (Kraemer, Luberto, O'Bryan, Mysinger, & Cotton, 2016). As previously mentioned, a study of survivors of the tsunami in Thailand in 2004 demonstrated that those with higher distress tolerance showed less emotional and physical signs consequences of stress related to the traumatic event (Piyasil et al., 2008). Mindfulness may serve as a valuable tool to prevent impairments after a disaster for those who regularly practice mindfulness prior to natural hazard or who show high trait mindfulness. For non-practitioners, mindfulness interventions delivered soon after the time of the natural hazard can serve to reduce subsequent effects after to prevent progression of harmful personal functioning outcomes reasonably within one's control. Continued practice after mindfulness interventions does not depend on external or costly tools, which helps in addressing disparities in access to care, making it accessible to those of various backgrounds and socioeconomic levels.

MINDFULNESS IN DISASTER SETTINGS

A small body of literature has examined mindfulness among those who have been exposed to a disaster. These studies have primarily focused on correlations with trait mindfulness. A few studies have investigated the application of mindfulness-based interventions among disaster survivors in efforts to reduce stress-related symptoms. The following section reviews both types of studies to evaluate the evidence base in support of providing mindfulness and meditation training for disaster survivors. Several studies in SEA have examined the various psychological benefits of mindfulness, but there is a paucity of research regarding mindfulness interventions for disaster relief.

TRAIT MINDFULNESS AND DISASTERS

There is some indication that trait mindfulness may convey resilience to post-disaster stress. A recent study found that mindfulness and attachment security were negatively correlated with PTSD symptoms among licensed, disaster mental health counselors who had responded to at least one disaster of any kind within the past five years (Pow & Cashwell, 2017). Trait mindfulness was prospectively associated with better adjustment in a longitudinal study of Israeli survivors of a large-scale forest fire which had resulted in forced evacuations, injury, and death (Nitzan-Assayag, Aderka, & Bernstein, 2015). Across three assessments conducted at one month, three months, and six months postdisaster, higher trait mindfulness predicted lower levels of distress in terms of depression, anxiety, negative affect, and PTSD symptoms.

There is some evidence that the beneficial relationship of trait mindfulness to posttraumatic stress symptoms may be weakened when disasters occur prior to trait mindfulness being fully developed. A recent study of Chinese adolescents who had been exposed to a tornado disaster found that trait mindfulness at six months postdisaster did not predict symptoms of posttraumatic stress at nine months postdisaster. However, higher posttraumatic stress symptoms at six months did predict lower trait mindfulness at nine months. The researchers suggested that the direction of this relationship provides evidence that among disaster-exposed adolescents, symptoms of posttraumatic stress may disrupt one's development of trait mindfulness. It was further suggested that posttraumatic stress symptoms may inhibit trait mindfulness through impairment of attentional capacities, a core component of mindfulness (An et al., 2018). A second study using a large sample of adolescents ($n = 435$) who had been exposed to the same tornado in Jiangsu province, China, found that those with higher dispositional mindfulness engaged in more adaptive strategies for coping with depression, when compared to those with lower levels of dispositional mindfulness (Xu, Ding, Goh, & An, 2018).

The degree to which trait mindfulness may serve as a protective factor may also depend on the severity of one's postdisaster symptoms. Among a sample of Norwegian tourists ($n = 25$) who were survivors of the 2004 tsunami in Thailand, a positive relationship was observed between symptoms of PTSD and trait mindfulness (Hagen, Lien, Hauff, & Heir, 2016). It may be important to note, as acknowledged by the study authors, that data were collected from the survivors several years after the disaster and more than half of those who were contacted to participate in the study declined to participate. Given the small sample size and the lack of psychopathology among the sample (Hagen et al., 2016), it may be that the protective nature of trait mindfulness observed in other studies may only be observed among those with more severe disaster-related psychological symptoms.

MINDFULNESS INTERVENTIONS FOR DISASTERS

A handful of recent studies provide preliminary evidence that teaching mindfulness practices to disaster-exposed individuals may reduce disaster-related psychopathology. A study of social service providers following Hurricane Sandy examined the impact of a postdisaster psychoeducational intervention which included a mindfulness component (Powell, Wegmann, & Shin, 2019). The intervention, Caregivers Journey of Hope, was comprised of four components, one of which involved instruction about resources and coping skills such as a facilitator-led mindfulness narrative and mindfulness breathing technique. Although the study authors noted several limitations to the study, such as lack of a comparative control group, the findings suggest that the intervention was effective at increasing knowledge of postdisaster reactions and also reducing current stress among participants (Powell et al., 2019). However, it is unclear how significant the role of mindfulness strategies alone was in the study, given the psychoeducational components that were also included.

Stronger evidence in support of mindfulness interventions for disaster survivors has utilized a standardized procedure. Waelde et al. (2008) conducted a pilot study examining the effects of a manualized meditation intervention for reducing psychological symptoms among mental health workers in New Orleans following Hurricane Katrina. All participants ($n = 20$) had been directly or indirectly exposed to injury or illness resulting from the hurricane. Participants attended a four-hour workshop and a home study program lasting 8 weeks, focusing on guided practice of meditation, mantra repetition, breathing awareness and breath-focused imagery, awareness and release of thoughts, feelings, and sensations. Anxiety, overall PTSD, and PTSD-related symptoms of re-experiencing and hyperarousal were significantly reduced after the 8-week intervention. This study also lacked a control group which severely limits the findings but does suggest such interventions are feasible in postdisaster settings (Waelde et al., 2008).

MINDFULNESS INTERVENTIONS IN SEA

Mindfulness interventions have also been applied specifically in the SEA region. Following a mass casualty tsunami in Sri Lanka, one study found support for a mindfulness-informed group therapy called Council for disaster-related trauma (Hoeberichts, 2012). The intervention was taught in workshops for disaster-relief counselors and incorporated various psychotherapeutic techniques. Although the treatment was based on a psychodynamic model, Hoeberichts (2012) described teaching guided meditation, mindfulness of bodily sensations, as well as attention to thoughts and feelings about trauma-related reactions. The researcher noted that incorporating these mindfulness strategies appeared to make the intervention culturally appropriate and therefore more effective, although no formal assessment measures were described.

A group-administered, mindfulness-informed psychological first aid (PFA) intervention was used for disaster response in the Philippines (Hechanova, Ramos, & Waelde, 2015). The culturally informed intervention was administered six weeks after the mass casualty 2013 super Typhoon Haiyan. The intervention taught local government employees who were disaster responders ($n = 98$) mindful breathing, the "letting go exercise," and sitting meditations several times throughout the workshop. Although there was not a control group for comparative analysis, the researchers found that the overall PFA intervention significantly improved coping ability among the participants from pre-test to post-test (Hechanova et al., 2015). Furthermore, qualitative report indicated most participants specifically named mindfulness strategies as being beneficial. Even though the intervention was based primarily on a PFA model, the study provides support for the use of a mindfulness-based intervention in the SEA region.

The results of a mindfulness training workshop that used a manualized mindfulness meditation and mantra program for mental health disaster responders 12 weeks after Typhoon Haiyan also supported the use of mindfulness interventions in SEA (Waelde, Hechanova, Ramos, Macia, & Moschetto, 2018). Participants attended a four-hour workshop followed by an eight-week home study program with weekly online surveys to report degree of personal practice of the mindfulness and mantra techniques. The training workshop was based on aneight-week mindfulness meditation and mantra program Inner Resources for Stress, adapted for use in a disaster setting. The workshop entailed guided practice of mindfulness techniques, such as mindful breathing, breath-focused imagery, a breath-focused mantra, and observing feelings, thoughts, and experiences without resistance or engagement ("letting go"). These strategies were then complemented by the home study program which included handouts and audio recordings of guided mindfulness and mantra practice. Among those who completed both pre-test and post-test measures following the workshop ($n = 62$), most participants found the training beneficial for their work as disaster responders as well as their ability to cope with their own stress of being mental health disaster workers. Among those who also completed the 8-week home study program ($n = 23$) minutes of mindfulness practice predicted pre/post improvements in depression, but not anxiety. The researchers found that those who practiced mindfulness more often reported fewer depressive symptoms compared to those who practiced mindfulness less often. It is unclear whether the mindfulness training reduced symptoms of trauma, as only one participant endorsed such symptoms and completed the full eight-week program (Waelde et al., 2018). Taken together, these studies suggest that mindfulness and mantra interventions may be especially suited to disaster contexts where skill building interventions are indicated across a broad proportion of the affect population to promote resilience and recovery.

CONCLUSION

Albeit limited, the initial studies on the use of mindfulness postdisaster suggest its potential for use in the Southeast Asian region. However, future research is needed to build upon the limitations of the present literature, and provide more robust evidence. Namely, there is a need for more robust studies including a manualized treatment and longitudinal studies. Furthermore, the use of control groups would allow for comparative analyses. A randomized controlled trial of mindfulness interventions with longer term follow ups to examine continuing effects for disaster victims would help remedy these shortcomings and may help inform disaster relief efforts.

Limitations notwithstanding, mindfulness interventions appear to be a promising response for SEA disaster survivors and responders. International agencies on disaster response have reiterated the importance of culturally appropriate interventions. Hechanova and Waelde (2017) suggested that in the SEA region, psychosocial interventions should consider variations in emotional expression, the salience of shame, interdependence common in collectivistic cultures, the importance of spirituality, and the variations in religious beliefs (Hechanova & Waelde, 2017). Because mindfulness does not require emotional expression, it may provide a means to addressing survivors' posttrauma symptoms in a non-threatening manner. Moreover, mindfulness teaches non-judgmental acceptance which addresses survivors' shame. Studies using mindfulness-informed group interventions also suggest that it can be used in group settings. Finally, although religions vary in SEA, a common denominator is the use of reflection and silence. Thus, mindfulness meditation may be mechanism that is familiar and acceptable to SEA survivors while being an inclusive approach to respond to their psychological needs.

REFERENCES

Aikens, K. A., Astin, J., Pelletier, K. R., Levanovich, K., Baase, C. M., Park, Y. Y., & Bodnar, C. M. (2014). Mindfulness goes to work: Impact of an online workplace intervention. *Journal of Occupational and Environmental Medicine*, *56*(7), 721–731. https://doi.org/10.1097/JOM.0000000000000209

Aldao, A., Nolen-Hoeksema, S., & Schweizer, S. (2010). Emotion-regulation strategies across psychopathology: A meta-analytic review. *Clinical Psychology Review*, *30*, 217–237. https://doi.org/10.1016/j.cpr.2009.11.004

An, Y., Yuan, G., Zhang, N., Xu, W., Liu, Z., & Zhou, F. (2018). Longitudinal cross-lagged relationships between mindfulness, posttraumatic stress symptoms, and posttraumatic growth in adolescents following the Yancheng tornado in China. *Psychiatry Research, 266*, 334–340. https://doi.org/10.1016/j.psychres.2018.03.034

Auty, K. M., Cope, A., & Liebling, A. (2017). A systematic review and meta-analysis of yoga and mindfulness Meditation in prison. *International Journal of Offender Therapy and Comparative Criminology, 61*(6), 689–710. https://doi.org/10.1177/0306624X15602514

Bellinger, D. B., DeCaro, M. S., & Ralston, P. A. (2015). Mindfulness, anxiety, and high-stakes mathematics performance in the laboratory and classroom. *Consciousness and Cognition, 37*, 123–132.

Bishop, S., Lau, M., Shapiro, S., Carlson, L., Anderson, N., Carmody, J., … Devins, G. (2004). Mindfulness: A proposed operational definition. *Clinical Psychology: Science and Practice, 11*, 230–241. https://doi.org/10.1093/clipsy.bph077

Bögels, S., & Restifo, K. (2014). Session 8: Are we there yet? A mindful path through parenting. In *Mindful parenting: A guide for mental health practitioners*. New York, NY: Springer Science+Business Media.

Briere, J., & Elliott, D. (2000). Prevalence, characteristics, and long-term sequelae of natural disaster exposure in the general population. *Journal of Traumatic Stress, 13*, 661–679. https://doi.org/10.1023/A:1007814301369

Brown, K. W., & Ryan, R. M. (2003). The benefits of being present: Mindfulness and its role in psychological well-being. *Journal of Personality and Social Psychology, 84*, 822–848. https://doi.org/10.1037/0022-3514.84.4.822

Center for Southeast Asian Studies. (2019). Retrieved from https://www.niu.edu/cseas/resources/countries.shtml

Chen, S. L.-S., Lee, C.-S., Yen, A. M.-F., Chen, H.-H., Chan, C.-C., Chiu, S. Y.-H., … Chang, J.-C. (2016). A 10-year follow-up study on suicidal mortality after 1999 Taiwan earthquake. *Journal of Psychiatric Research, 79*, 42–49. https://doi.org/10.1016/j.jpsychires.2016.04.007

Chiesa, A., & Serretti, A. (2009). Mindfulness-based stress reduction for stress management in healthy people: A review and meta-analysis. *The Journal of Alternative and Complementary Medicine, 15*, 593–600.

Descilo, T., Vedamurtachar, A., Gerbarg, P., Nagaraja, D., Gangadhar, B. N., Damodaran, B., … Brown, R. P. (2010). Effects of a yoga breath intervention alone and in combination with an exposure therapy for post-traumatic stress disorder and depression in survivors of the 2004 South-East Asia tsunami. *Acta Psychiatrica Scandinavica, 121*, 289–300. https://doi.org/10.1111/j.1600-0447.2009.01466.x

Dyregrov, A., & Regel, S. (2012). Early interventions following exposure to traumatic events: Implications for practice from recent research. *Journal of Loss and Trauma, 17*, 271–291. https://doi.org/10.1080/15325024.2011.616832

Grecucci, A., Pappaianni, E., Siugzdaite, R., Theuninck, A., & Job, R. (2015). Mindful emotion regulation: Exploring the neurocognitive mechanisms behind mindfulness. *BioMed Research International, 2015*, 1–9. https://doi.org/10.1155/2015/670724

Guendelman, S., Medeiros, S., & Rampes, H. (2017). Mindfulness and emotion regulation: Insights from neurobiological, psychological, and clinical studies. *Frontiers in Psychology, 8*, 1–23. https://doi.org/10.3389/fpsyg.2017.00220

Hagen, C., Lien, L., Hauff, E., & Heir, T. (2016). Is mindfulness protective against PTSD? A neurocognitive study of 25 Tsunami disaster survivors. *Journal of Negative Results in BioMedicine, 15*, 13. https://doi.org/10.1186/s12952-016-0056-x

Hechanova, R. M., Ramos, P. A. P., & Waelde, L. C. (2015). Group-based mindfulness-informed psychological first aid after Typhoon Haiyan. *Disaster Prevention and Management: An International Journal, 24*, 610–618. https://doi.org/10.1108/DPM-01-2015-0015

Hechanova, Ma. R. M., & Waelde, L. C. (2017). The influence of culture on disaster mental health and psychosocial support interventions in Southeast Asia. *Mental Health, Religion & Culture, 20*, 31–44. https://doi.org/10.1080/13674676.2017.1322048

Hoeberichts, J. H. (2012). Teaching Council in Sri Lanka: A post disaster, culturally sensitive and spiritual model of group process. *Journal of Religion and Health, 51*, 390–401. https://doi.org/10.1007/s10943-010-9358-3

Jia, Z., Tian, W., He, X., Liu, W., Jin, C., & Ding, H. (2010). Mental health and quality of life survey among child survivors of the 2008 Sichuan earthquake. *Quality of Life Research, 19*, 1381–1391. https://doi.org/10.1007/s11136-010-9703-8

Kiken, L. G., & Shook, N. J. (2012). Mindfulness and emotional distress: The role of negatively biased cognition. *Personality and Individual Differences, 52*(]]), 329–333. https://doi.org/10.1016/j.paid.2011.10.031

Kim, Y. W., Lee, S.-H., Choi, T. K., Suh, S. Y., Kim, B., Kim, C. M., … Yook, K.-H. (2009). Effectiveness of mindfulness-based cognitive therapy as an adjuvant to pharmacotherapy in patients with panic disorder or generalized anxiety disorder. *Depression and Anxiety, 26*, 601–606. https://doi.org/10.1002/da.20552

Kraemer, K. M., Luberto, C. M., O'Bryan, E. M., Mysinger, E., & Cotton, S. (2016). Mind–body skills training to improve distress tolerance in medical students: A pilot study. *Teaching and Learning in Medicine, 28*, 219–228. https://doi.org/10.1080/10401334.2016.1146605

Lee, W. K., & Bang, H. J. (2010). The effects of mindfulness-based group intervention on the mental health of middle-aged Korean women in community. *Stress and Health, 26*, 341–348. https://doi.org/10.1002/smi.1303

Mackenzie, C. S., Poulin, P. A., & Seidman-Carlson, R. (2006). A brief mindfulness-based stress reduction intervention for nurses and nurse aides. *Applied Nursing Research, 19*, 105–109. https://doi.org/10.1016/j.apnr.2005.08.002

Mahoney, C. T., Segal, D. L., & Coolidge, F. L. (2015). Anxiety sensitivity, experiential avoidance, and mindfulness among younger and older adults: Age differences in risk factors for anxiety symptoms. *The International Journal of Aging and Human Development, 81*, 217–240. https://doi.org/10.1177/0091415015621309

Myers, S., & Wells, A. (2015). Early trauma, negative affect, and anxious attachment: The role of metacognition. *Anxiety, Stress, & Coping, 28*, 634–649. https://doi.org/10.1080/10615806.2015.1009832

Netz, Y., & Lidor, R. (2003). Mood alterations in mindful versus aerobic exercise modes. *The Journal of Psychology, 137*, 405–419. https://doi.org/10.1080/00223980309600624

Nitzan-Assayag, Y., Aderka, I. M., & Bernstein, A. (2015). Dispositional mindfulness in trauma recovery: Prospective relations and mediating mechanisms. *Journal of Anxiety Disorders, 36*, 25–32. https://doi.org/10.1016/j.janxdis.2015.07.008

North, C. S., & Pfefferbaum, B. (2013). Mental health response to community disasters: A systematic review. *JAMA, 310*, 507. https://doi.org/10.1001/jama.2013.107799

Osofsky, H., Osofsky, J., Kronenberg, M., Brennan, A., & Hansel, T. C. (2009). Posttraumatic stress symptoms in children after Hurricane Katrina: Predicting the need for mental health services. *American Journal of Orthopsychiatry, 79*, 212–220. https://doi.org/10.1037/a0016179

Phang, C. K., Mukhtar, F., Ibrahim, N., Keng, S.-L., & Mohd. Sidik, S. (2015). Effects of a brief mindfulness-based intervention program for stress management among medical students: The Mindful-Gym randomized controlled study. *Advances in Health Sciences Education, 20*, 1115–1134. https://doi.org/10.1007/s10459-015-9591-3

Piyasil, V., Ketumarn, P., Prubrukarn, R., Pacharakaew, S., Dumrongphol, H., Rungsri, S., … Chatchavalitsakul, W. (2008). *Psychiatric disorders in children at one year after the tsunami disaster in Thailand. Journal of the Medical Association of Thailand, 91*(Suppl 3), S15–S20.

Pow, A. M., & Cashwell, C. S. (2017). Posttraumatic Stress Disorder and emotion-focused coping among disaster mental health counselors. *Journal of Counseling & Development, 95*, 322–331. https://doi.org/10.1002/jcad.12146

Powell, T. M., Wegmann, K. M., & Shin, O. J. (2019). Stress and coping in social service providers after Superstorm Sandy: An examination of a postdisaster psychoeducational intervention. *Traumatology, 25*, 96–103. https://doi.org/10.1037/trm0000189

Tang, B., Liu, X., Liu, Y., Xue, C., & Zhang, L. (2014). A meta-analysis of risk factors for depression in adults and children after natural disasters. *BMC Public Health, 14*(1). https://doi.org/10.1186/1471-2458-14-623

Teut, M., Roesner, E. J., Ortiz, M., Reese, F., Binting, S., Roll, S., … Brinkhaus, B. (2013). Mindful walking in psychologically distressed individuals: A randomized controlled trial. *Evidence-Based Complementary and Alternative Medicine*, 1–7. https://doi.org/10.1155/2013/489856

Thompson, B. L., & Waltz, J. (2010). Mindfulness and experiential avoidance as predictors of posttraumatic stress disorder avoidance symptom severity. *Journal of Anxiety Disorders*, *24*(4), 409–15. https://doi.org/10.1016/j.janxdis.2010.02.005

Udomratn, P. (2008). Mental health and the psychosocial consequences of natural disasters in Asia. *International Review of Psychiatry*, *20*, 441–444. https://doi.org/10.1080/09540260802397487

UNISDR and World Bank. (2010). Synthesis report on ten ASEAN countries disaster risks assessment. Retrieved from https://www.unisdr.org/files/18872_asean.pdf

Vujanovic, A., Bonn-Miller, M., Potter, C., Marshall, E., & Zvolensky, M. (2011). An evaluation of the relation between distress tolerance and posttraumatic stress within a trauma-exposed sample. *Journal of Psychopathology and Behavioral Assessment*, *33*, 129–135. https://doi.org/10.1007/s10862-010-9209-2

Waelde, L. C. (2015). Mindfulness and meditation for trauma-related dissociation. In V. Follette, J. Briere, J. Hopper, D. Rozelle, & D. Rome (Eds.), *Mindfulness-oriented interventions for trauma: Integrating contemplative practices* (pp. 301–313). New York, NY: Guildford Press.

Waelde, L. C., Hechanova, Ma. R. M., Ramos, P. A. P., Macia, K. S., & Moschetto, J. M. (2017). Mindfulness and mantra training for disaster mental health workers in the Philippines. *Mindfulness*. Advance online publication. https://doi.org/10.1007/s12671-017-0855-2

Waelde, L. C., Hechanova, R. M., Ramos, P. A. P., Macia, K. S., & Moschetto, J. M. (2018). Mindfulness and mantra training for disaster mental health workers in the Philippines. *Mindfulness*, *9*, 1181–1190. https://doi.org/10.1007/s12671-017-0855-2

Waelde, L. C., Meyer, H., Thompson, J. M., Thompson, L., & Gallagher-Thompson, D. (2017). Randomized controlled trial of Inner Resources meditation for family dementia caregivers. *Journal of Clinical Psychology*, Advance online publication. http://doi.org/10.1002/jclp.22470

Waelde, L., Panting, A., & Heise, A. (2019). Mindfulness in the peacebuilding process. In M. Njoku, L. Jason, & B. Johnson, *The psychology of peace promotion* (p. 25). New York, NY: Springer.

Waelde, L. C., & Thompson, J. M. (2016). Traditional and secular views of psychotherapeutic applications of mindfulness and meditation. In M. A. West (Ed.), *The psychology of meditation: Research and practice* (pp. 119–152). Oxford, U.K.: Oxford University Press.

Waelde, L. C., Uddo, M., Marquett, R., Ropelato, M., Freightman, S., Pardo, A., & Salazar, J. (2008). A pilot study of meditation for mental health workers following Hurricane Katrina. *Journal of Traumatic Stress*, *21*, 497–500. https://doi.org/10.1002/jts.20365

Weinstein, N., Brown, K. W., & Ryan, R. M. (2009). A multi-method examination of the effects of mindfulness on stress attribution, coping, and emotional well-being. *Journal of Research in Personality*, *43*, 374–385. https://doi.org/10.1016/j.jrp.2008.12.008

World Health Organization Regional Office for South-East Asia. (2019). Retrieved from http://www.searo.who.int/countries/en/

Xu, W., Ding, X., Goh, P. H., & An, Y. (2018). Dispositional mindfulness moderates the relationship between depression and posttraumatic growth in Chinese adolescents following a tornado. *Personality and Individual Differences*, *127*, 15–21. https://doi.org/10.1016/j.paid.2018.01.032

Yook, K., Lee, S.-H., Ryu, M., Kim, K.-H., Choi, T. K., Suh, S. Y., … Kim, M.-J. (2008). Usefulness of Mindfulness-Based Cognitive Therapy for treating insomnia in patients with anxiety disorders: A pilot study. *The Journal of Nervous and Mental Disease*, *196*, 501–503. https://doi.org/10.1097/NMD.0b013e31817762ac

CHAPTER 7

ART THERAPY AS A DISASTER RESPONSE IN SOUTHEAST ASIA: STATE OF THE ART

J. Sedfrey S. Santiago

ABSTRACT

The occurrences of natural hazards in Southeast Asia have become not only more frequent but their severity has also intensified. More than 200,000 persons perished from the 2004 tsunami resulting from the 9.1 magnitude earthquake in coastal Sumatra, Indonesia, which was the third strongest since 1900. A record-breaking typhoon internationally named Haiyan (locally known as Yolanda) ravaged Central Philippines in 2013. Failure of the communities, as well as the countries, to cope with the hazards has led to disasters that have compelled them to seek external assistance, especially internationally. In the 2004 and 2013 disasters referred to, one specific form of assistance was the conduct of art therapy especially for children. This chapter surveys the literature and examines why and how art therapy has been used in disasters in the region, and which areas either need further exploration or research.

Keywords: Art therapy; disasters; trauma; posttraumatic stress disorder (PTSD); posttraumatic growth (PTG); recovery

The occurrences of natural hazards in Southeast Asia have become not only more frequent but their severity has also intensified. Hydro-meteorological events like typhoons and seismic activities have adversely affected many countries in the region (Samphantharak, 2019). The fact that a considerable part

Resistance, Resilience, and Recovery from Disasters: Perspectives from Southeast Asia
Community, Environment and Disaster Risk Management, Volume 21, 109–116

ISSN: 2040-7262/doi:10.1108/S2040-726220200000021007

of Southeast Asia lies along the Pacific Ring of Fire makes the region highly vulnerable to seismic-related disasters (Hechanova & Waelde, 2017; Parr, 2015; Samphantharak, 2019).

More than 200,000 persons perished from the 9.1 magnitude earthquake in coastal Sumatra, Indonesia, the third strongest recorded since 1900 (Tsunami of 2004 Fast Facts, 2013). A typhoon internationally named Haiyan (locally known as Yolanda) also ravaged Central Philippines in 2013. With its maximum sustained winds peaking at 250 kilometers per hour-speed, Haiyan is so far the strongest typhoon ever to hit land (Santiago, Manuela, Tan, Sañez, & Tong, 2016). Approximately 6,300 people died because of Typhoon Haiyan, with 1,602 more missing (Bueza, 2016).

The International Federation of Red Cross and Red Crescent Societies stated the concept of disaster in a formula: (vulnerability + hazard)/capacity = disaster (What is a disaster?, ifrc.org., n.d.). The IFRC formulation considers disaster from the ability of the affected community or sector to cope. The severity of a disaster is dependent on capacity to cope relative to the affected community's vulnerability and the hazard's magnitude combined. The foregoing characterization reflects a commonly used definition of "disaster"; that is, an event that causes widespread losses and impact, seriously disrupting a community's functioning, and exceeding its ability to cope using the community's own resources (Yasuyuki, Bhattcharyay, & Tomoaki, 2011; Republic Act No. 10121). There is thus one common and important element among the definitions, namely, the ability of the affected community, using its own resources, to respond to the hazard and cope with damages.

THE IMPACT OF DISASTERS

There has been much written about the psychological impact of disasters. A specific example is Sri Lanka which was struck by tsunami following the December 2014 earthquake in Sumatra. Children who personally saw the horrific demise of their loved ones were unduly exposed to a high risk of posttraumatic stress disorder (PTSD) (Chilcote, 2007). PTSD is a common postdisaster condition (Hechanova & Waelde, 2017), with trauma being described as the feeling of terror, vulnerability, or distress by a person as a result of experiencing a situation or event that poses threat of serious injury or even death (American Psychiatric Association, 2000).

However, most survivors' PTSD symptoms are temporary, and the initial number of PTSD manifestations tends to decrease as time elapses (Hechanova & Waelde, 2017). The decreasing manifestations notwithstanding, there remains a need to address trauma because of the serious threats posed to the normal psychological functioning of survivors (Parr, 2015). This has led to the call for mental health interventions as early as possible following a disaster. In particular, there are those who have suggested the need to focus on children and adolescents as they are more vulnerable than adults to the traumatic events of disasters (Math et al., 2008).

DELINEATING ART THERAPY

One way to help the child cope with trauma is using art as a medium of expression (Parr, 2015). Art therapy plays a role in postdisaster situations because survivors often cannot verbalize their experience and creative arts is a means of expressing "the nonverbal aspects of the traumatic imprint in order to safely access the memory" (Mohr, 2014, p. 156). Parr (2015, p. 138) also wrote that expressive arts is an "alternative means for people to talk about themselves and their experiences in an authentic and effective way."

An illustration of an art therapy intervention is One Heart ExPress, an initiative led by academics and artists from the University of the Philippines that has used expressive arts like theater, visual, performance, literature, and music as an intervention in Palo, Leyte, and Quinapondan, Eastern Samar, and Bohol (Santiago et al., 2016). However, the project called PLATR ("Phuket Life Access Tracking") that was launched in 2006 after the 2004 Indian Ocean tsunami disaster (Nobuto, 2016) cannot be considered as art therapy despite its intended benefit to the community for several reasons: (a) Japanese artists from outside the community created art in the community as part of their residency in Phuket and (b) the workshops conducted by the artists for students focused on the arts and culture rather than the disaster and possible trauma that the students experienced.

Mohr (2014) asserted that the effectiveness of art-making as a tool in helping survivors cope with trauma is connected to the importance of physical, somatic, and body-oriented treatment. Moreover, art's therapeutic properties have been proven effective not only with children and adolescents but also with adults (Parr, 2015). In some parts of Asia though, the use of art therapy is relatively new. Recently, Ahmed and Siddiqi (2006) wrote that "(i)n developing countries, especially in south Asia, art therapy is almost non-existent" (p. 529). Within a broader context, it has also been observed that "(t)he sparse literature" on mental health and psychosocial support "shows that the science of disaster response is in its infancy in SEA" (Hechanova & Waelde, 2017, p. 40).

THE VALUE OF ART THERAPY

Art therapy is subsumed within play therapy, a type of intervention that has long been used for children with mental health issues (Ahmed & Siddiqi, 2006). The "use of drawing with traumatized children enables them to express emotions and verbalize experiences more effectively than talking alone" (Chilcote, 2007, p. 157). Expressive arts therefore helps survivors comprehend what they have gone through and "allows them to reframe their stories in a way that has meaning to them" – constructing, deconstructing, or reconstructing – and retelling their stories – whether individually, or as a member of family or community (Parr, 2015, p. 140).

Chilcote (2007) further explained:

> Stronach-Buschel (1990) found that it is helpful for a child to visualize the traumatic experience before putting it into words. Art therapy enables children to reproduce the trauma symbolically, thereby controlling emerging memories and integrating experiences without becoming overwhelmed. Art therapy helps to empower traumatized children by rebuilding their fractured

> sense of competency and control (Stronach-Buschel, 1990). Rankin and Taucher (2003) utilized art therapy to enable traumatized children to express their current emotional state, to narrate the traumatic event, to process feelings, and to integrate the trauma into their life histories (p. 157).

Using her own experience with post-tsunami traumatized children in Sri Lanka, Chilcote (2007) stated that through art therapy, "the children were able to regain emotional control that was shattered with the tsunami and to commemorate their loss" (p. 161). In addition, children in the Andaman and Nicobar Islands (Indonesia), who also survived the December 2004 earthquake and tsunami, were asked to describe their drawings. The children's articulation of their art proved valuable to the therapists as it enabled the therapists to identify which children were more severely affected by the disaster, and therefore needed more support and intervention (Math et al., 2008).

OF RISKS AND DESYNCHRONIZATIONS

One important aspect of psychosocial interventions, including art therapy, is that the culture of the community engaged with influences implementation (Hechanova & Waelde, 2017). Chilcote (2007) proposed that art therapists must be flexible enough to adapt to different cultural norms. She cited that Western concepts of privacy and confidentiality do not fit in a "community-oriented culture where bedrooms do not have doors but rather curtains, and where family relationships are deeply rooted in one another's lives" (p. 157).

Based on the foregoing, we suggest that materials used in art therapy should also be well thought-out and not haphazardly introduced to the intervention kit. For instance, if dolls are used, it must be remembered that dolls are not only children's toys, "they are (also) pieces of material culture that embody and convey social beliefs and values" (Sweet, 2014, para. 1). The community's religious beliefs, like aniconism, should be thoughtfully considered and must be respected.

Another imperative is the management and coordination of intervention programs to prevent the recurrence of what happened to some survivors of Typhoon Washi in 2011. Although well-meaning, the programs of different international and local agencies and NGOs, using varying techniques and approaches, were uncoordinated and unsupervised (Santiago et al, 2016). "Participants in post-disaster focus group discussions reported that they 'were asked to go through different healing sessions one after the other, which left them feeling aggrieved instead'" (Santiago et al, 2016, p. 651). One way to address this issue is for groups introducing art therapy to establish partnerships with institutions in the community like local government units and religious organizations, who can help in managing and scheduling intervention programs. Community-based institutions can also facilitate the implementation of art therapy projects because they are presumed to be more familiar with the community being engaged. Specifically, in the case of religious organizations, the religious grounding of the affected community may play "a major role in helping them come to terms with the destruction" (Math et al., 2008, p. 37).

THE BANGLOS FISHERFOLKS STORY AS EXCEPTION[1]

Not all art-related projects introduced to the community are mainly therapeutic. The outreach program of self-taught Filipino sculptor Rey Paz Contreras, for instance, started more than thirty years ago (Santiago, 2015). The program aims to teach sculpting skills using discarded wood and other materials. The program focuses on marginalized communities and the sculptures are meant to be sold as part of the community members' income-generating activities. Contreras started within the community that he lived in – a poor community in Tondo, Manila located beside the railroad track – using discarded *traviesa* or railroad ties or sleepers as basic material.

Contreras' introduction to the Banglos fisherfolks[2] of General Nakar in Quezon Province was through a corporation that has been helping the fisherfolks as part of its corporate social responsibility program. The community is not a stranger to typhoons as General Nakar is a usual path for typhoons coming from the Pacific Ocean. In 2004, the worst of these weather disturbances forced the fisherfolks to permanently abandon their residences because the waters that submerged their community never receded. Also, their fishing areas were choked by wood debris washed down by the typhoon from the mountains. Farming was not an alternative livelihood because their farms were similarly covered with heaps of dead wood. Despite the need for money, the fisherfolks initially resisted the idea of learning how to sculpt as an alternative livelihood because they only knew how to fish and plant; and the fisherfolk loathed to touch wood from the mountains that had caused them death and destruction (Santiago, 2015).

After realizing that sculpting discarded wood was not as difficult as it seemed, the fisherfolks relented, and began to create art as they continued to rebuild their lives. Subsequently, through the help of corporate sponsors, the art works were shown and sold in exhibits in prestige venues like the Cultural Center of the Philippines. Proceeds from exhibit sale were used for the education of the fisherfolks' children or purchase of necessities like motorcycles. Although not one fisherfolk has become a full-time artist, this new experience has had an effect beyond economics. The majority of the fisherfolks interviewed themselves have articulated on the knowledge (*kaalaman*) and capacity (*kakayanan*) they have acquired as a result of the art workshops. This articulation indicates a sense of empowerment. The fisherfolks' new-found confidence has enabled them not only to address their economic needs but also to give back. And this giving back, which is one of the values imparted by Contreras, has taken two forms: (a) mentoring others on sculpting such as the Capiz folks who survived Typhoon Haiyan as a means of livelihood and (b) donating part of their exhibit sales proceeds to other disaster survivors.

One incidental effect of the art-making experience was that sculpting has proved therapeutic for the fisherfolk. This effect, albeit incidental to the main objective, affirms the observation "that simple expression through art helps children to come to terms with their painful memories" (Ahmed & Siddiqi, 2006, p. 529) except that in this case, those who have benefitted from the process are adults.

ARTICULATING THE GAPS

Considering the infancy of mental health intervention programs in the region, including art therapy, there is a need to develop the literature on existing practices that can help improve the extant and be a basis for innovations. Parr (2015) enumerated some of the possible themes or topics, such as: (a) in-depth assessment of the innovative family-focused approach, (b) inquiry into other expressive arts modalities such as dance and materials like clay and how they can facilitate survivors' communication of their traumatic experiences, and (c) inquiry into the types of trauma experienced by those who have survived man-made disasters like armed conflicts, especially to find out what specific types of expressions they best respond to.

Similarly, there is limited literature on group art therapy in the region. However, based on earlier studies in other regions, this approach has been shown to be "an effective method to help disaster survivors deal with a multitude of issues" (Murphy 2014, p. 12). Group therapy's effectiveness arises from the fact that it "can involve large segments of an affected community, agency, or organization and reach more people in a time-efficient manner" (Murphy, 2014, p. 16; see also Math et al., 2008).

Proceeding from the Banglos fisherfolks' experience; future study could address the therapeutic effect of art-making on adults, even when art is introduced to the community mainly for economic reasons. This effort would also help develop research and literature on art therapy intervention programs for adults, and how adults are specifically affected. Studies on how art therapy can help and strengthen adults, especially parents and first responders, are important because these are the people who are naturally expected to care of children, shielding them as much as possible from danger and trauma.

The Silver Lining.

Postdisaster trauma and how best to address it need not be the sole focus of further studies. Beyond the negative consequence of trauma is posttraumatic growth (PTG). Mohr (2014) conducted research among Peruvian teens who were undergoing art therapy three years after they survived an earthquake in the cities of Ica and Pisco. Mohr stated:

> Tedeschi and Calhoun (2004) defined posttraumatic growth as a positive experience of change that may occur as a result of a struggle with difficult challenges or trauma. Such struggles may bring heightened appreciation of meaning in interpersonal relationships, greater sense of personal strength, shifts in priorities, increased appreciation for small life events, and a richer interior life in the existential or spiritual sense (p. 155).

Clarifying that her study did not seek to establish a "cause–effect relationship between relief work in art therapy and posttraumatic growth," Mohr (2014) concluded:

> The youth underscored the important role of the arts in bringing comfort and joy to others, coupled with direct verbal validation of one another's lived experiences where necessary and possible. The honoring, enriched relatedness, increased empathy, and nuanced insights that these youth imaged forth and described may be important indicators of posttraumatic growth. (p. 161).

CONCLUSION

This review indicates that studies of the utilization and utility of art therapy to address children's trauma after disasters in Southeast Asia exist. There is ample room, however, for more research. Evaluation of previous art therapy intervention in affected communities, and a study including PTG theme in Southeast Asia are possible gaps that can be addressed. This future work could develop and refine existing intervention programs with the aim of strengthening the programs to achieve two proposed objectives: (a) maximization of the efficiency of art therapy programs and avoidance of any unfavorable effects, albeit unintended; and (b) identification of factors that may enhance transformation of trauma's negativity to promotion of postdisaster recovery and growth.

NOTES

1. This section is substantially based on a paper entitled "Community-Based Mentoring On Sustainable Art: Lessons For The Academe?". Presented by the author in the International Association of Jesuit Business Schools (IAJBS) 21st Annual World Forum held at the Catholic University of Uruguay, Montevideo, Uruguay on July 19–21, 2015.
2. Banglos is a combination of *Bang*on (Rise) and *Kilos* (Act).

REFERENCES

Ahmed, S. H., & Siddiqi, M. N. (2006). Healing through art therapy in disaster settings. *Lancet*, *368*, S28–S29. Retrieved from www.thelancet.com/pdfs/journals/lancet/PIIS0140673606699169.pdf

American Psychiatric Association. (2000). *Diagnostic and statistical manual of mental disorders* (4th ed., text rev.). Washington, DC: Author.

Bueza, M. (2016, November 7). In numbers: 3 years after Super Typhoon Yolanda. Retrieved from https://www.rappler.com/newsbreak/iq/151549-in-numbers-3-years-after-super-typhoon-yolanda-haiyan

Chilcote, R. L. M. A. (2007). Art therapy with child tsunami survivors in Sri Lanka. *Art Therapy*, *24*(4), 156–162. doi: 10.1080/07421656.2007.10129475

Hechanova, R., & Waelde, L. (2017). The influence of culture on disaster mental health and psychosocial support interventions in Southeast Asia. *Mental Health, Religion & Culture*, *20*, 31–44. doi: 10.1080/13674676.2017.1322048

Math, S. B., Tandon, S., Girimaji, S. C., Benegal V, Kumar U, Hamza A, Jangam K, Nagaraja D. (2008). Psychological impact of the tsunami on children and adolescents from the Andaman and Nicobar Islands. *Primary Care Companion to the Journal of Clinical Psychiatry*, *10*, 31–37. doi: 10.4088/pcc.v10n0106

Mohr, E. (2014). Posttraumatic growth in youth survivors of a disaster: An arts-based research project. *Journal of the American Art Therapy Association*, *31*, 155–162. https://doi.org/10.1080/07421656.2015.963487

Murphy, C. F. (2014). *Post-disaster group art therapy treatment for children*. Master's thesis, Loyola Marymount University, Los Angeles, CA, USA. Retrieved from http://digitalcommons.lmu.edu/etd/55

Nobuto, K. (2016). *Community reconstruction through cultural identity regeneration tourism and creativity in Phuket, Thailand*. Unpublished manuscript. Osaka: Department of Asian Culture and Urbanism, Osaka City University.

Parr, R.A. (2015). The use of expressive arts in making sense of traumatic experiences. *Philippine Journal of Psychology*, *48*, 133–159. Retrieved from www.pap.org.ph/sites/default/files/pdf/PJP1502_Final_6Parr.pdf

Republic Act No. 10121 (approved May 27, 2010) also known as the Philippine Disaster Risk Reduction and Management Act. of 2010.

Samphantharak, K. (2019). Natural disaster and economic development in Southeast Asia. Forthcoming, Economic Development of Southeast Asia (2019) MIT Press. Retrieved from https://papers.ssrn.com/sol3/papers.cfm?abstract_id=3388396

Santiago, J. S. (2015, April 6–13). Of driftwood and fishermen sculptors. Rey Paz Contreras comes to General Nakar, Quezon. *Philippines Graphic Magazine*, *25*(45 & 46), 28–30.

Santiago, J. S., Manuela, W. S., Tan, M. L., Sañez, S. K., & Tong, A. Z. (2016). Of timelines and timeliness: Lessons from Typhoon Haiyan in early disaster response. *Disasters*, *40*, 644–667. https://doi.org/10.1111/disa.12178

Sweet, E. (2014). Dolls. Oxford Bibliographies. Retrieved from https://www.oxfordbibliographies.com/view/document/obo-9780199791231/obo-9780199791231-0155.xml?rskey=UmrO4F&result=13&q=sweet#firstMatch.

Tsunami of 2004 Fast Facts. (2013, August 23). Retrieved from https://edition.cnn.com/2013/08/23/world/tsunami-of-2004-fast-facts/index.html.

What is a disaster? (n.d.). Retrieved from www.ifrc.org/en/what-we-do/disaster-management/about-disasters/what-is-a-disaster/.

Yasuyuki, S., Bhattcharyay, R., & Tomoaki, K. (2011). Aggregate impacts of natural and man-made disasters: A quantitative comparison (RIETI Discussion Paper Series 11-E-023). Retrieved from https://www.rieti.go.jp/jp/publications/dp/11e023.pdf

CHAPTER 8

MENTAL HEALTH AND PSYCHOSOCIAL SUPPORT SERVICES FOR DISASTER RESPONDERS IN SOUTHEAST ASIA

Johnrev Guilaran and Hong An Nguyen

ABSTRACT

Disaster responders play a crucial role in providing aid to individuals and communities following catastrophic events. Being tasked to protect and preserve life and property, these groups of professionals are constantly exposed to various hazards, which puts them at risk of negative mental health consequences. This chapter describes and discusses these mental health effects and interventions for disaster responders in Southeast Asia. The chapter defines who the disaster responders are in Southeast Asian countries. Drawing from the literature, this chapter enumerates the various positive and negative psychological consequences of disaster response, and the risk and protective factors associated with disaster response work. This chapter also describes the different interventions, such as psychological first aid and psychotherapy, following the Inter-agency Standing Committee (IASC) (2007) guidelines on conducting mental health and psychosocial support services (MHPSS), and focusing on the Southeast Asian context. This chapter ends with a discussion of the different challenges of providing MHPSS in Southeast Asia and with some recommendations on how to improve the delivery of these services and the mental health of disaster responders in general.

Keywords: Disasters; disaster responders; Southeast Asia; MHPSS; mental health; psychosocial support services

Resistance, Resilience, and Recovery from Disasters: Perspectives from Southeast Asia
Community, Environment and Disaster Risk Management, Volume 21, 117–142

ISSN: 2040-7262/doi:10.1108/S2040-726220200000021008

Disasters are overwhelming situations or events that require national or international assistance (Guha-Sapir, Hoyois, & Below, 2016). The Centre for Research on the Epidemiology of Disasters (EM-DAT) defines disasters as events that (1) have 10 or more people killed, (2) have 100 or more people affected, (3) require a declaration of a state of emergency, or (4) a call for international assistance. As disasters challenge local capacities to cope, the role of disaster responders is most salient in parts of the world where the occurrence of these events is high, such as in Asia. Some of the most vulnerable countries are found in the Southeast Asian region (Below & Wallemacq, 2018), which is composed of Brunei, Cambodia, East Timor (Timor-Leste), Indonesia, Laos, Malaysia, Myanmar, Philippines, Singapore, Thailand, and Vietnam. However, despite its disaster vulnerability, research on Southeast Asian disaster mental health represents a small percentage of the general disaster mental health literature (Hechanova & Waelde, 2017). Studies on the general population (e.g., Norris et al., 2002) and on disaster responders (e.g., Guilaran, de Terte, Kaniasty, & Stephens, 2018) showed that most disaster mental health studies were conducted in the United States despite the fact that majority of disasters occurred in Asia. This underrepresentation was exacerbated by the disproportionately small number of mental health studies on disaster workers (see Norris & Elrod, 2006; Norris et al., 2002).

This chapter focuses on disaster responders in Southeast Asia – individuals who put themselves on the line to protect and preserve life, property, and the environment (Prati & Pietrantoni, 2010a) in the aftermath of disasters and other catastrophic events in Southeast Asia. Because of the scarcity of literature on Southeast Asian responders, ideas in this chapter were also drawn from the general literature and from the literature on other Asian responders and contextualized in Southeast Asia.

DISASTER RESPONDERS IN SOUTHEAST ASIA

Disasters affect people in different ways. Some people will go through the experience unscathed while others may be traumatized by the exposure. Although there is an abundance of studies on the effects of trauma, disasters are complex collective experiences that influence a similarly complex set of mental health consequences. A majority of individuals exposed to these gruesome and overwhelming occurrences are unwilling participants, yet there are those who put themselves forward to assist individuals in need of support in the aftermath of experiencing these events (Brooks, Dunn, Amlôt, Greenberg, & Rubin, 2016). As a general observation, exposure to disasters, including disaster response work, puts individuals at risk of experiencing psychological reactions (McFarlane & Norris, 2006). This observation is validated in Southeast Asian responders (e.g., Chan, Tang, Hall, Yip, & Maggay, 2016). However, the extent to which disasters affect the psychological outcomes of disaster responders is complex and is affected by several factors, which are described later in this chapter.

Generally, disaster responders may be categorized as traditional or non-traditional depending on their mandate (Pietrzak et al., 2014). Traditional responders are those who are formally or officially tasked to respond in the aftermath of these critical events. Examples of these are firefighters, police, military, and people working in the emergency medical services. On the other hand, non-traditional responders are those who are not mandated to respond after disasters, but who perform as responders due to circumstance. For example, primary school teachers in the Philippines usually find themselves providing psychosocial support services to students in the aftermath of disasters even if they are not mandated to do so. In Vietnam, it is not rare to see people from the communities volunteering to provide aid after disasters. They do so by joining humanitarian/charitable aid organizations or by organizing their own groups to provide material and emotional support to disaster survivors. It is also not uncommon for the public works sector to be involved in post-disaster clean up, reconstruction, and rehabilitation, even though they are not specifically a disaster response sector. Particularly in low-to-middle-income Southeast Asian countries, there is an increasing participation of non-traditional responders in carrying out post-disaster response (Bush, 2015).

In Southeast Asia, disaster responders can be residents of the disaster-hit areas (making them primary survivors as well), may come from the unaffected parts of the affected country, or from other countries as part of international aid. In cases of large-scale disasters and events involving mass casualties, such as the 2004 Boxing Day Tsunami and the 2013 Typhoon Haiyan, disaster responders from overseas are common. Although there are government-mandated disaster responders, such as members of the civil defense force, non-government organizations play a huge role in the disaster response activities in the Region.

DISASTER STRESSORS

Disasters are stressful events that challenge the coping capacity, not only of the community, but also of the individual experiencing it. Experiencing disasters often lead to a wide range of psychological consequences. The most potent predictors of severity of post-disaster psychological outcomes are the proximity, duration, and intensity of disaster exposure (Benedek, Fullerton, & Ursano, 2007). Although exposure to disasters puts people at risk of psychological distress, the majority of exposed individuals will only experience transient effects (Norris & Elrod, 2006). A significant number of people will exhibit acute reactions in the immediate aftermath, but would normalize a few weeks later (Dunn, Brooks, & Greenberg, 2015). However, a small portion of the exposed population will develop serious mental health concerns.

In terms of exposure, the disaster-affected population may be classified into three groups: primary, secondary, and tertiary (De Soir, 2006). Primary survivors are those who are affected by the disaster first-hand. This group would include people who were provided medical assistance or those who were rescued from the event. Secondary survivors are comprised of individuals with significant others (e.g., family members, very close friends) who were injured, killed, or significantly affected by the disaster. Finally, tertiary survivors are composed of individuals who respond to these events (e.g., first responders, journalists). Tertiary survivors have not experienced the disaster first-hand, but have experienced the trauma vicariously (e.g., seeing its effects, working with survivors, and knowing their suffering).

This, of course, is not a straightforward classification when applied to disaster responders. Though these professionals are usually exposed to the disaster vicariously, they also may experience the disaster directly. Others experience offshoots of the disaster while on disaster response work, as in the case of experiencing devastating aftershocks while responding in the aftermath of an earthquake. Furthermore, they may also have family members or very close friends who were affected by the disaster. In this sense, disaster responders may encompass all three levels of exposure (Alexander & Klein, 2009), making them a category of exposed individuals with special needs. Responders may also be primary or secondary survivors these events, and this may affect not only themselves, but also the people they are responding to (Dunn et al., 2015; Naturale & Pulido, 2012).

The conditions in which disaster responders operate also add to the adversities that they encounter. Exposure to dead bodies is one of the most stressful experiences of disaster work, especially when it involves dead children (Alexander & Klein, 2009). In addition to exposure to its psychologically disturbing elements, disaster responders face risks, either perceived or actual, such as exposure to toxic chemicals and communicable diseases. The roles and tasks that responders perform may also put them at risk of experiencing reactions. For example, tasks like providing psychosocial support services and relief aid were linked to post-traumatic stress symptoms in Indonesian disaster workers (Thormar et al., 2013). Working in disaster sites means enduring inadequate sleeping arrangements, insufficient equipment, and fatigue (Alexander & Klein, 2009). Disaster responders who are unfamiliar with the culture and social terrain of the deployment area may also experience culture shock (Puig & Glynn, 2004).

This complexity in the influence of disaster exposure on psychological outcomes is exacerbated by the socio-political climate in some Southeast Asian countries. In the Philippines, for example, disaster workers may be caught in the middle of political conflict (Ladrido-Ignacio, 2011), such as when politicians use disaster response as a vehicle to gain political influence. They may also find themselves entangled in the problems of their respective organizations (Guilaran, 2013), which adds to the cumulative stress. In Vietnam, disaster aid was found to entail corruption and abuse of relief funds (Nguyen, 2016). Occurrences like these can negatively impact social trust and volunteer activities (Woo, Cheng, Ho, & Horowitz, 2015), which may significantly affect the morale of disaster responders.

PSYCHOLOGICAL CONSEQUENCES OF DISASTER RESPONSE

When one responds to a disaster, exposure to potentially disturbing sensory experiences such as mutilated bodies, injuries, people's suffering, and highly distressed individuals may be a necessary component of the job (Alexander & Klein, 2009).

Such exposure may trigger a wide range of emotional and behavioral reactions. In order to provide proper and adequate mental health and psychosocial support services (MHPSS) for disaster responders, it is imperative to understand the range of mental health concerns that they may experience as a consequence of working in these perilous situations. The psychological consequences of disaster exposure are widely documented. These consequences range from normative to clinical reactions, with most people experiencing mild distress. Few people will experience moderate symptoms and a much smaller group will probably experience serious psychiatric conditions (Benedek et al., 2007). The following are some of the psychological reactions related to disaster work.

Vicarious traumatization (VT) happens when the disaster responder exhibits intrusion, avoidance, hypervigilance, and dissociation symptoms, which are similar to that of posttraumatic stress disorder (PTSD) (Naturale & Pulido, 2012). The major difference between VT and PTSD is the exposure to trauma. Whereas PTSD may develop as a consequence of direct exposure to disasters, VT is a result of exposure to the suffering of those who experienced the disaster first-hand (Palm, Polusny, & Follette, 2004). It is argued that professionals whose work involves empathic interaction with survivors are more prone to developing vicarious trauma reactions. Within the range of disaster first responders, emergency health service workers and disaster mental health workers may be the most at risk of these reactions.

Sometimes used interchangeably with compassion fatigue (CF), secondary traumatic stress (STS) happens when the disaster worker experiences symptoms similar to the symptoms of the people they are helping, as a consequence of being indirectly exposed to the trauma these people have experienced or are experiencing (Creamer & Liddle, 2005). However, CF and STS are different concepts. CF is more closely associated with the helping profession (e.g., counselors, psychotherapists) while STS is more appropriately used to refer to traumatic reactions in more diverse populations, including disaster responders (Ludick & Figley, 2017). Both CF and STS may be experienced by the disaster responder as consequences of prolonged exposure to other people's suffering and heightened empathic response. On one hand, prolonged exposure to suffering and traumatic memories contributes to the experience of STS. A major symptom of STS is when the helping professional avoids the client and the reminders of the client's trauma, while having re-experiencing symptoms (Adams, Boscarino, & Figley, 2006). A study of disaster relief workers in the Philippines pointed to feeling tired and frustrated as manifestations of STS (Guilaran, 2013). On the other hand, CF occurs when these types of exposures influence empathic responses. CF happens when a professional experiences reduction of caring or empathic capacity as a result of knowing about the traumatic experiences of a person. In Cambodia, due to lack of resource and support, teachers working with traumatized children were found to lack self-care and were at risk of CF (Wyatt, Hoban, & Macfarlane, 2018).

Burnout is another common reaction to chronic emotional and interpersonal work-related stress, characterized by exhaustion, cynicism, and inefficiency (Maslach, Schaufeli, & Leiter, 2001). Mental health service providers are most at risk of experiencing this reaction (Cieslak et al., 2014). Too much workload may result in the experience of burnout. This is particularly an issue in some parts of Southeast Asia. In the Philippines, for example, disaster relief workers report having to do their regular development and humanitarian work on top of their disaster relief work during disaster deployment, which may lead to cumulative stress (Guilaran, 2013).

In addition to these reactions, a small percentage of disaster responders may experience PTSD. Although PTSD is one of the more widely observed clinical outcomes associated with disaster work, it is not the most common single diagnosis in disaster responders (Alexander & Klein, 2009). A meta-analysis of 28 studies suggested that about 10% of rescue workers may satisfy the criteria for PTSD (Berger et al., 2012). A study of 237 Typhoon Haiyan responders in the Philippines documented significant relationships between disaster-related stressors and general psychological distress and PTSD (Chan et al., 2016). The prevalence of PTSD and PTSD symptoms ranged from 8.6% to 57.3% after the 2004 Boxing Day Tsunami (Udomratn, 2008). Disaster responders were also found to be at risk of experiencing depressive and anxiety disorders (Norris et al., 2002; North et al., 2002).

Aside from these conditions, other reactions to disaster work may include bereavement (Alexander & Klein, 2009) and trauma-related and/or survivor's guilt (Tam, Pang, Lam, & Chiu, 2004). Changes in coping be behavior may also be observed, such as increase in alcohol consumption (Alexander & Klein, 2009), although this is thought to be an intensification of old coping patterns, reinforced by organizational culture, rather than a start of a new one. Other disaster responders may cope by suppressing emotions, which may be reinforced by the macho culture in many traditional responder organizations (Villaflor, Villaret, & Hechanova, 2019). Avoidant coping behavior may also be observed in these professionals (Guilaran, 2013). Disaster responders may undergo changes in their worldview. Disaster work may influence the disaster responders' sense of justice and their assumptions about the world. In addition, it may also shape their expression of emotions such as shame, anger, and guilt (Alexander & Klein, 2009).

There is bias toward research on the negative outcomes of disaster work. In a meta-analysis on the effects of social support in disaster responders, a large percentage of studies were found to focus on negative psychological consequences (Benedek et al., 2007; Guilaran et al., 2018). However, recent developments in the field have shifted the attention from psychopathology to resilience and well-being (Alexander & Klein, 2009). For example, frontline health workers during the SARS outbreak in Hong Kong experienced some positive outcomes, such as rethinking of their life's priorities, their values, and better social relationships (Tam et al., 2004). People who work in disasters may also experience post-traumatic growth (Tedeschi & Calhoun, 2004), a positive psychological change attributed to going through very difficult circumstances in life, such as disasters and other traumatic events. For example, although some conflict between family members were observed after the 2004 Tsunami, improvement in family relationships and positive changes in family values were reported (Lindgaard, Iglebaek, & Jensen, 2009).

FACTORS ASSOCIATED WITH POST-DISASTER PSYCHOLOGICAL OUTCOMES

There are a number of factors associated with the severity of psychological outcomes following exposure to disaster. Some demographic risk factors for the development of PTSD in disaster responders include being younger, being single, and having prior trauma exposure (Kleim & Westphal, 2011). Although females are usually observed in the general literature to have a higher risk of psychological distress (Norris & Elrod, 2006), some evidence in Southeast Asia point to being male as a risk factor. In contrast to mainstream literature, male Indonesian disaster workers were found to have more depressive symptoms than females (Thormar et al., 2013), which was attributed to cultural factors such as lack of acceptance of expressed emotions which fuels internalization. Evidence also points to masculine ideology as a barrier to help-seeking among Filipino emergency responders (Villaflor et al., 2019), putting them at higher risk for developing more serious psychological conditions.

Differences in reactions were found between different types of responders, although more research is still needed to support these observations, particularly in Southeast Asia. For example, non-traditional responders were found to be more vulnerable to PTSD than their traditional counterparts (Bromet et al., 2016). This difference is attributed to factors such as training, personality characteristics, type and degree of exposure, and level of peer support (Alexander & Klein, 2009). However, volunteers may also report more positive experience out of disaster work than traditional responders – attributed to the meaningfulness of the endeavor rather than performing the tasks as part of a mandate (Alexander & Klein, 2009).

Certain psychological and social resources are also associated with resilient post-disaster trajectories. Self-efficacy is a protective factor and lack of which puts the responders at risk of psychological distress (Benight & Bandura, 2004). For example, being injured and consequently, becoming dependent on others (i.e., being a burden) was found to be associated with increased risk in developing serious mental illness among disaster responders in the Philippines (Chan et al., 2016). On the other hand, optimism was found to be associated with resilient outcomes after adversity in police officers (de Terte, Stephens, & Huddleston, 2014). Hardiness is another psychological protective trait associated with positive post-disaster outcomes in disaster responders. People who are considered hardy have a better sense of control over events, frame difficulties as challenges rather than as threats, and find meaning in the whole experience (Alexander & Klein, 2009).

Spirituality and religion were also found to help disaster responders cope with the stress brought about by deployment. While there is considerable diversity of religions practiced in Southeast Asia, the significant role that spirituality plays in the lives of its people is crosscutting. Spirituality influences how people make sense of disasters, affects how they cope, and provides additional layer of social support. However, religion may also be a deterrent to help-seeking and may shape unhealthy attribution of the damage caused by disasters (Hechanova & Waelde, 2017).

Social support is one of the most reliable protective factors against psychological distress (Brewin, Andrews, & Valentine, 2000; Ozer, Best, Lipsey, & Weiss, 2003). Social support refers to behaviors and social interactions that provide actual assistance and embed people in a network of loving and caring relationships (Hobfoll & Stokes, 1988). The effect of social support has been conventionally framed in two ways, either as a direct effect or as a stress-buffer effect (Kaniasty & Norris, 2009). The direct effects model suggested that social support influences mental health outcomes, regardless of conditions of stress. For example, increase in social support was found to be associated with lower levels of posttraumatic stress symptoms in Swedish survivors of the 2004 Southeast Asian Tsunami, regardless of the level of exposure (Arnberg, Hultman, Michel, & Lundin, 2012). Alternatively, the stress-buffering model suggested that social support itself does not contribute to positive psychological outcomes but in the face of adversity it mitigates the effects of stress, thereby reducing its impact on health and well-being. Another way of looking at the moderating effect of social support is that it can magnify psychological outcomes in the face of trauma, referred to in the literature as "reverse buffering effect" (Prati & Pietrantoni, 2010b) or "outcome enhancement effect" (Guilaran et al., 2018).

Although social support is generally associated with good psychological outcomes, it is a finite resource and has been observed to deteriorate with time, especially in the context of disasters (Kaniasty & Norris, 1995, 2004; Norris & Kaniasty, 1996). This is important to note, as resource loss is associated with the experience of psychological distress (Hobfoll, 1989). It is also important to distinguish the various facets of social support, which have differential effects on the mental health outcomes of disaster responders. Kaniasty and Norris (2009) discussed the distinctions among received social support (receipt of actual support), perceived social support (evaluation of support quality and availability), and social embeddedness (social integration). Of the three facets of social support, perceived social support has been shown to be more consistently associated with positive psychological outcomes in disaster responders – having quality and readily available support is seen to influence how disaster responders view risks and helps them identify effective ways of coping (Prati & Pietrantoni, 2010b). However, the effectiveness of receiving actual support is influenced by several factors. The support-matching hypothesis (Cutrona & Russell, 1990) posited that the effectiveness of social support is hugely influenced by the match between the support needed and the support provided. The social support effectiveness hypothesis (Rini & Dunkel Schetter, 2010) further proposed that received social support effectiveness depends on the match between the quality and quantity of support needed and provided, the difficulty in obtaining support or the spontaneity of offering it, the quality of its provision, and the consequences of receiving it (Rini, Dunkel Schetter, Hobel, Glynn, & Sandman, 2006).

Furthermore, the dynamics of social support mobilization and utilization are strongly influenced by culture (Chen, Kim, Mojaverian, & Morling, 2012; H. S. Kim, Sherman, & Taylor, 2008; Taylor et al., 2004), such as the preference for the type of support and the comfort in seeking for support. Social support dynamics are also shaped by culture-related socioeconomic factors, such as the level of pre-disaster resources associated with certain ethnic groups, which influences support-seeking and support provision in the aftermath of disasters (Kaniasty & Norris, 2000). Having stated this, more research is needed in understanding the dynamics of social support in Southeast Asia, particularly in disaster responders.

IMPLICATIONS FOR SOUTHEAST ASIA

Disaster risk is formulated as the interaction of exposure to hazards (e.g., typhoons, earthquakes) and vulnerability, which has the dimensions of susceptibility (e.g., state of infrastructure), coping capacity (e.g., preparedness), and adaptive capacity (e.g., long-term adaptive mechanisms) (Welle & Birkmann, 2015). These dimensions of exposure and vulnerability may also be used to assess the extent to which disaster responders in Southeast Asia, and disaster responders in general, may experience psychological distress and other negative psychological consequences related to disaster response work. Following this formulation, higher risk of negative psychological consequence may be experienced by those with greater disaster exposure and with a greater degree of vulnerability. However, the wide variation of exposure and vulnerability characteristics in Southeast Asia poses a challenge in contextualizing disaster risk in the Region. This challenge of contextualization also applies to the disaster responders working in the different countries in this geographic area.

Some Southeast Asian countries have higher degrees of exposure to hazards than others. For example, the Philippines, being situated in the typhoon belt, experiences a larger share of typhoons and storms than the rest of the Region. Southeast Asian countries in the Pacific Ring of Fire (e.g., Philippines, Indonesia) also experience more earthquakes and other seismic-related hazards than parts outside the area (e.g., Thailand). Areas surrounding the Chao Phraya river system and the Mekong Delta have more chronic exposure to extreme flooding than the other parts of the Region. There are cross-country variations in the degree of exposure, such as differences in the magnitude or strength of the hazard experienced or differences in the frequency of hazard occurrence, or both. For example, countries like the Philippines experience more typhoons, and also stronger typhoons, than the rest of Southeast Asia. With all things being equal, higher degree of exposure to hazards means more intense effects on the population of the affected areas, and consequently, more intense effects experienced by the responders, either first-hand or vicariously.

Along with the variation in the level of hazard exposure, the degree of vulnerability also varies across the Region. Some parts of Southeast Asia have higher susceptibility (i.e., susceptibility, coping capacity, and adaptive capacity) to the adverse effects of hazard exposure than others. For example, people in countries like the Philippines and Vietnam have better means of acquiring disaster-related information than people in Myanmar, which may predict how citizens in these countries may prepare and cope with the effects of hazards (Lai, Chib, & Ling, 2018). Disaster responders who have higher vulnerability to the negative (psychological) effects of hazard exposure are usually those who come from countries with fewer resources to build capacities for coping (both in the short- and long-term) and for reducing susceptibility. Susceptibility is the likelihood that one may suffer from harm as a result of the exposure. The variation of vulnerability of responders in Southeast Asia may also be attributed to the differences in disaster-related policies across the countries in the Region. Disaster responders who have no/have substandard equipment or those with inadequate training in disaster response may be susceptible to the adverse effects of hazards. The lack of protective equipment may expose them to danger while responding to hazards. Lack of proper training may also influence self-efficacy, which is a strong predictor of posttraumatic psychological reactions such as symptoms of PTSD and depression. Furthermore, disaster responders who belong to organizations that lack the infrastructure to help them deal with psychosocial concerns may also jeopardize the disaster responders' ability to cope with the immediate effects and adapt to the more long-term effects of hazard exposure. The heterogeneity of exposure and vulnerability contexts in the different areas of the Southeast Asian Region should be considered in the planning and implementation of MHPSS for disaster responders in the area.

MHPSS INTERVENTION

The Inter-agency Standing Committee (IASC, 2007) presented a set of guidelines with the following six core principles in providing MHPSS. First, MHPSS is a human right, and equity and fairness must be exercised at all times. Second, the conduct of MHPSS must maximize participation of the recipients of the intervention. Third, it must be ensured that the conduct of MHPSS does not cause harm. Fourth, MHPSS activities should optimize the strengths and resources of its recipients. Fifth, MHPSS are not standalone activities. It is recognized that mental health outcomes are influenced by other elements of a larger system, such as basic services, health care, housing, and education; hence, MHPSS should be integrated with the provision of these services. Sixth, disasters affect people in different ways, which creates a wide range of needs. As these needs can be addressed in various ways, the IASC proposed a hierarchy of supportive actions, which may also apply to MHPSS activities for disaster responders.

Basic Services and Security

The foundation of every MHPSS program for disaster responders is ensuring that their basic needs are met. People responding to disasters should have adequate food and shelter. This is challenging during deployment, especially in the immediate aftermath of disasters, where resources are scarce. However, it is necessary that disaster responders have the basic services to ensure that they are able to perform their tasks effectively. Response organizations should make sure that there is adequate food for its members. They should also be housed in a safe space, away from hazards.

Part of addressing the basic services and security of disaster responders is by making sure they have adequate financial remuneration (i.e., salary) and other benefits. In addition, as noted earlier, many disaster responders in Southeast Asia may also be disaster survivors. In circumstances like this, disaster responders should first take care of themselves and their families before responding to other survivors. Organizations should also make sure that the families of disaster responders are safe and secure while they are being deployed.

Community and Family Supports

As mentioned earlier, having social support buffers the negative effects of disaster exposure. The second tier in the IASC hierarchy addresses the need for community and family support. Most disaster responders will survive post-disaster chaos for as long as there is ease of access of support from various sources. Social support is an important resource, especially in times of disasters. Emotional support, for example, is associated with lower levels of PTSD symptoms (Gabert-Quillen et al., 2012). Having social support also allows one to have access to other forms of support (e.g., money). It allows individuals to extend their resources beyond the personal domain (Hobfoll, Freedy, Lane, & Geller, 1990). Various sources of support offer differential benefits to disaster responders. For example, friends and family offer invaluable support, by providing them with space for adjustment after deployment (Alvarez & Hunt, 2005). MHPSS activities can center around strengthening the disaster responders' social networks, particularly with significant supportive sources, such as creating peer support groups or fortifying existing social networks.

The organization can play a huge role in this MHPSS tier. Organizational support can come in various forms. An important support that response organizations can provide is through skills training and psychoeducation. Self-efficacy is a significant predictor of reduction of psychological distress (Benight & Bandura, 2004; Prati, Pietrantoni, & Cicognani, 2010). Disaster responders who have better sense of control over the situation are found to experience fewer distress symptoms. Instilling skills and knowledge is one way to increase one's sense of self-efficacy. Although they may not come across as conventional psychological interventions, training disaster responders on skills that are crucial to their tasks, such as basic life support skills, may increase their sense of control. This also underscores the importance of preparedness before disaster responders go on deployment. It is imperative that they know the tasks to be done, and they are aware of what to expect during and after deployment. This also decreases the risk of burnout.

In addition, organizations can support disaster responders by creating an environment that facilitates positive outcomes. There are organizational factors associated with reducing the negative impact of disaster work. Good organization; clearly defined duties; attention to personal needs, particularly the physical; good teamwork; and, appreciation of one's work contribute to the buffering of the negative consequences of disaster work (Alexander & Klein, 2009). A clear organizational structure does not only allow for better disaster response, but it also minimizes the confusion among disaster responders (Panyayong & Pengjuntr, 2006), which may cause psychological distress. The same thing can be said of clearly defined roles and responsibilities. During disaster response, it is also easy for personal and professional time to be enmeshed. Organizations need to ensure that there is a clear delineation between time spent at work and time spent for oneself.

Organizations can help buffer the negative effects of disaster work. For example, identification with the deceased is found to be a potent risk factor for psychological distress and other serious mental health issues (Alexander & Klein, 2009). Disaster response organizations can put preventive measures in place. Strategies like distancing, not learning the names of the deceased individuals, wearing proper protective clothing, use of masking oils to deal with offensive odor were found to help reduce distress on the part of the responder (Alexander & Klein, 2009). By providing these practical types of support, organizations minimize the experience of psychological distress among its members. Some organizations can be very creative in helping their members deal with the stressful nature of disaster work. Some organizations would set up areas (e.g., an empty container van) where disaster workers can just shout to vent frustration or to let off steam.

Furthermore, organizations also need to ensure that there is a good balance between working hours and off hours. Regular supervision, peer-group supervision, trainings, and workday self-care plans also protect disaster responders from psychological distress (Naturale & Pulido, 2012). Disaster responders may be deployed from a number of days to several months. It is the responsibility of the organization to ensure that the disaster responders have ample breaks in between deployment (e.g., break after every two weeks). This will provide the responders the time to rest and recuperate.

Organizations can support responders through preparation. Evidence drawn from military studies show that those who know what to expect in the combat zone have better post-deployment mental health outcomes (e.g., Franz et al., 2013). It was also observed that disaster relief workers who know what to expect on site experience lower levels of distress (Guilaran, 2013). Some strategies for preparing disaster workers include psychoeducation on the nature of the disaster operating environment, simulations, survival training, comprehensive briefing, capacitating leaders, and developing supportive networks among the workers (Paton, 1996).

Focused, Non-specialized Supports

The third tier in the IASC intervention hierarchy focuses on the smaller subgroup of disaster responders who may need individually tailored support. At the heart of this MHPSS layer is the use of psychological first aid (PFA), which was designed to be offered in the immediate aftermath of disasters, and other interventions for use in the medium to longer term after a disaster.

Psychological first aid.

PFA has gained ground in the last decade as a means of providing post-disaster intervention. Although PFA has been mostly utilized for disaster survivors, it has also been found to be effective for disaster responders (Alexander & Klein, 2009). PFA is a set of guiding principles in dealing with the immediate aftermath of exposure to disasters and other emergencies. The course of action suggested by PFA aims toward reducing psychological distress and increasing adaptability and post-trauma functioning (Ruzek et al., 2007). Preference on the use of PFA in immediate post-disaster intervention is due to its consistency with evidence on posttraumatic and resilience research, applicability in a wide range of settings and across different developmental stages, and its flexibility and cultural sensitivity (Pekevski, 2013; Vernberg et al., 2008).

PFA is based on the five basic empirically supported principles of psychological intervention: promotion of (1) sense of safety, (2) calm, (3) sense of self- and community-efficacy, (4) connectedness, and (5) hope (Hobfoll et al., 2007). These guiding principles informed the creation of eight core actions in providing post-disaster psychological support (Vernberg et al., 2008). The first core action is the initiation of contact and engagement. At the heart of this process is the identification of disaster responders who need immediate help. When PFA providers enter the scene, they try to get to know the disaster responders and use their informed judgment to identify who needs or wants contact, and eventually who needs further attention (Ruzek et al., 2007). Every intervention process must begin with a respectful, non-intrusive initial contact. This process establishes rapport and sets the tone of the entire helping process. It is also important to start this core action with knowledge of, and sensitivity toward, cultural norms and practices. In fact, the IASC (2007) suggested that intervention providers should be sensitive to cultural elements such as the observance of acceptable and/or comfortable social distance, expression of emotions, and other culturally sanctioned behaviors. This core action also lays the foundation for the second core action, which is the establishment of safety and comfort.

Exposure to traumatic events such as disasters may trigger fight or flight responses even emergency and disaster workers (Iserson et al., 2008) who are constantly exposed to such events; hence, it is crucial to establish a sense of safety and comfort. Disasters challenge the sense of security and make salient one's sense of mortality. Exposure to disasters puts one in a state of self-preservation, but at the same time, one may experience emotional and biological responses such as freeze reactions, overwhelming fear, and dissociation (Shalev, 2000). With these considered, the PFA provider may facilitate a safe and comfortable atmosphere for disaster responders – one that conveys physical and psychological safety (Pekevski, 2013).

The third core action involves stabilizing, which involves techniques that ground exposed individuals who are experiencing disorientation, hyperarousal, or emotional numbing (Vernberg et al., 2008). This stage is geared toward providing a sense of calm (Pekevski, 2013), which cascades to better psychological and social functioning. Grounding techniques are usually associated with this stage.

Information gathering is the next core action. This stage aims at knowing the disaster responder's needs, immediate concerns, current reactions, and other information which are vital to the conduct of the other PFA core actions

(Vernberg et al., 2008). This core action aims at identifying (1) the problems that require immediate attention; (2) the disaster responders who may need future intervention; and, (3) the issues that need to be addressed in the other PFA core actions. In other words, this stage is important for creating a hierarchy of issues to be addressed. In addition, this core action provides information in performing a psychological triage. These are very important in the context of disasters, which, more often than not, is characterized by problems of scarcity of time and human resources.

One of the strengths of PFA is that it redefines the whole concept of psychological intervention. This is exemplified by the fifth core action, which is the provision of practical assistance. This core action, of course, is guided by the information gathered in the previous core action. Evidence has shown that those who have lost resources in the aftermath of disasters are usually those who will experience the most distress (Norris & Kaniasty, 1996). However, provision of practical support for disaster responders should be handled carefully. In some instances, providing practical support may have unintended consequences (Deelstra et al., 2003). For example, when practical support is delivered in a way that affects the disaster responder's sense of efficacy (e.g., a sense that they are helpless and they cannot deal with their own problems), it may result in psychological distress. This is where the importance of information gathering comes in. The PFA provider must know the kind of practical assistance needed by the disaster responder so that the provided support matches the support needed. Furthermore, the PFA provider should gather information on the preferred manner in which practical support is to be provided, as certain disaster responder groups have existing organizational cultures that may not value the idea of being helped or supported (i.e., being weak and vulnerable). It also matters who provides the support (Halbesleben, 2006). Nevertheless, providing practical support is as important as providing the more conventional psychological support such as emotional support or advice.

In the event that PFA providers cannot give the support needed, they may refer the disaster responder to those who can. This is the essence of the sixth core action, which is the linking of the affected individual to other sources of support. This core action is linked to evidence pointing to social connectedness as a reliable protective factor against psychological distress in the aftermath of disasters (Hobfoll et al., 2007). This does not only link the disaster responder with other sources of support, but also links them with people who have similar experiences. As disaster responders usually work with teams, social connectedness appears to be built within the system. However, it must be noted that people cope in different ways. Other disaster responders may feel the need to avoid other people in the immediate aftermath of the exposure. Bringing together people with similar experiences may also lead to them simultaneously letting off steam, which may create a "pressure cooker effect." This unintended product of social support happens when individuals share about their traumatic experiences with one another, resulting in the increase of the level of psychological distress due to overexposure (Prati & Pietrantoni, 2010b). Nevertheless, evidence shows that connecting disaster responders with the right sources of support was positively associated with good mental health outcomes (e.g., Haslam & Mallon, 2003). In this respect, having sensitive and reliable peers/teammates helps (Guilaran, 2013), as they provide readily available and appropriate support, especially on-site.

The next core action is the provision of information on stress reactions and coping. A widely used form of this core action is psychoeducation. This core action provides the disaster responder information on adaptive and maladaptive reactions, and ways on how to properly deal with psychological distress. A crucial part of this core action is the establishment of routines for disaster responders who have just returned from deployment. This stage also involves colleagues, non-work friends, and family members. These people also need to be informed of the psychological reactions that are to be expected post-deployment, and the proper way to deal with these reactions (Vernberg et al., 2008).

Finally, the eighth core action involves linking disaster responders with other services. This follows through the intervention established in the PFA and ensures that needs are addressed by other individuals or agencies that are capable of addressing the disaster responder's concerns. For example, the need for prolonged leaves of absence may be addressed by the human resource department of the response organization; hence, the PFA provider may liaise between the department and the disaster responder. It also normalizes the act of help-seeking. An important reason why disaster responders may not receive help is because they hesitate to ask for help (see Villaflor et al., 2019). By linking them with other resources and/or other people who may link them to these resources, it sends a message of help-seeking as a normative behavior.

It must be emphasized that PFA was designed to be carried out by non-psychologists or non-specialists; hence, it does not assume that the PFA provider can perform specialized treatment such as psychotherapy. In fact, this is one of the strongest points of the PFA. In regions like the Southeast Asia where human resources are scarce, PFA optimizes the availability of lay individuals to perform "first aid" until the specialists arrive or become available. Although PFA was originally designed for one-on-one intervention, certain circumstances, like the lack of human and other resources (Landoy, Hechanova, Ramos, & Kintanar, 2015) in Southeast Asia led to its adaption as a group process. This appears to work with the collectivistic culture orientation of most of the Region.

Though these core actions should be followed in order, some situations may not require the performance of all steps. For instance, some response organizations have in-house mental health providers who may already have established rapport with the disaster responder prior to deployment and may not need a thorough contact and engagement phase (as this already has been established). Some disaster responders may also not need stabilizing, so this stage may be skipped. The important point in conducting PFA is that these are guiding principles that can be adjusted depending on the need and the context. Furthermore, it must be emphasized at this stage that the different core actions are ordered in such a way that the previous core action facilitates the conduct of the next. For example, gathering information is easier when the disaster responder is stabilized, and for obvious reasons, gathering information is easier when there has been an effective establishment of contact.

Although more research is needed to test the efficiency and effectiveness of PFA, this method is being endorsed because of its preference for non-psychiatric labeling, its focus on addressing immediate post-disaster needs, and its facilitation of coping (Alexander & Klein, 2009). It can also be provided by non-specialists, which makes it very accessible in the aftermath of disasters, where the need overwhelms availability of human resources. However, since PFA is relatively new, and while its principles are based on evidence, there is still a need for studies evaluating its effectiveness (Pekevski, 2013).

Medium- to long-term interventions

PFA is a short-term intervention and there are also medium-term interventions for disaster responders. An example of this are the *Katatagan* modules (Hechanova, Waelde, & Ramos, 2016) developed for disaster survivors in the Philippines, which may also be used for disaster responders. Designed as a tool for resilience-building, *Katatagan* has six modules that target specific resilience dimensions (Hechanova, Ramos, & Waelde, 2015). The *Kalakasan* module focuses on the discovery and cultivation of one's strengths. The *Katawan* module focuses on the management of physical reactions. The *Kalooban* module focuses on the management of thoughts and emotions. The *Kalutasan at Kaagapay* module promotes problem-solving and support-seeking. The *Kapaki-pakinabang na Gawain* module centers on the promotion of positive activities. The *Kinabukasan* module facilitates moving forward. These modules were designed using principles of cognitive behavioral therapy (CBT) and positive psychology (Hechanova et al., 2015) and were designed as a group-based intervention. These modules have been pilot-tested on survivors of the 2013 Typhoon Haiyan in the Philippines (Flores et al., 2014; Hechanova, Waelde, Docena, et al., 2016).

Another example is Inner Resources for Stress (Waelde, 2014) which is a mindfulness, meditation, and mantra intervention designed to help disaster responder acquire skills for self-care. A study of Inner Resources for Stress with disaster workers in New Orleans, Louisiana following Hurricane Katrina found pre- to post-intervention reductions in PTSD and anxiety symptoms, with greater amounts of daily practice of the techniques associated with more improvement of these symptoms (Waelde et al., 2008). Inner Resources for Stress was also tested in the Philippines following Super Typhoon Haiyan. Because disaster workers are so often disaster-exposed, this study examined whether the intervention was feasible and useful for disaster-exposed responders. Results indicated that both disaster-exposed and non-exposed participants have high expectancies that the training would help with their own self-care and benefit their work with survivors. Growth curve analyses showed that the disaster-exposed group had a slightly flatter trajectory in the number of minutes per week of personal practice of the techniques, suggesting that they might benefit from more support during the early weeks of the eight-week intervention. However, both groups of disaster responders had significant increases in their minutes of practice of the techniques per week, indicating that the intervention was feasible for disaster-exposed workers who were exposed to ongoing stresses of the disaster zone.

Aside from interventions provided by another person/group, disaster responders may also practice self-care. The idea is simple: you cannot give what you do not have. Disaster responders need to make sure that they are in a healthy state, both physically and psychologically, for them to effectively perform their tasks. At the core of self-care is taking care one's physical health. This may be achieved through regular exercise, proper diet, routine medical appointments, and adequate sleep. The nature of disaster work may easily disrupt these activities, so it is important that disaster responders find time to do these activities on schedule.

Self-care also means taking care of one's mental health. It is easy to feel isolated during and after deployment, so it is important to constantly connect with people within and outside the organization. Talking about one's concerns may lessen the psychological burden and ease the distress. As noted earlier, finding meaning in one's work also optimizes the positive gains of disaster exposure. Having a hobby or an enjoyable activity outside disaster work also provides a welcome distraction. Consistent with studies in Southeast Asia, cultivating one's spirituality may be beneficial for the disaster responder. This may be done through having quiet time or through meditation and mindfulness exercises. Finally, disaster responders should never forget to find humor, as it reduces stress and tension.

Although it is imperative for disaster responders to take care of themselves, it must be emphasized that self-care should never take the place of an organizational prevention and intervention system. Organizations that deploy first responders are liable to the health and well-being of their members.

Specialized Services

A much smaller group of disaster responders may require more specialized forms of intervention. These responders have been affected by disaster work in a way that significantly impairs their daily functioning. This MHPSS layer includes psychotherapy and other psychiatric forms of intervention. Psychopharmacology is one of those specialized treatment procedures used to address serious conditions such as PTSD and acute stress disorder (Benedek et al., 2007). However, this section will focus on the psychological treatments.

One of the most popular forms of post-trauma psychological interventions for survivors and responders alike is psychological debriefing, although its effectiveness is not supported by empirical evidence (North & Pfefferbaum, 2013). However, there are evidence-based psychological interventions for dealing with effects of traumas such as disaster exposure. For example, psychotherapies such as CBTs, narrative exposure therapy (NET), and eye movement desensitization and reprocessing (EMDR) are some interventions for PTSD supported by evidence (Haugen, Evces, & Weiss, 2012).

The effectiveness of CBT in PTSD treatment, especially trauma-focused CBT, is widely accepted (National Institute for Health & Clinical, 2005). Several meta-analyses and RCT studies show that CBT significantly improves PTSD symptoms in Southeast Asia's traumatized population, in both individual (Bryant et al., 2011; Hinton et al., 2004, 2005; Palic & Elklit, 2011) and group intervention formats

(Pityaratstian et al., 2015). It was found to be an effective intervention for the survivors of the 2004 Boxing Day Tsunami (Udomratn, 2008). Its techniques involve psychoeducation, exposure, cognitive restructuring, and teaching coping and relaxation skills. In addition, several modifications on the modality made it more culturally sensitive. These modifications include the use of metaphors, the focus on community's beliefs, and communication in a culturally acceptable manner (Otto & Hinton, 2006). For instance, the use of the image of the lotus blooming in the wind is used to depict flexibility among Vietnamese trauma survivors (Hinton et al., 2004).

Another group of evidence-based modalities that are increasingly becoming popular are the narrative approaches. Of particular interest in more recent literatures is the NET. NET focuses on the client's comprehensive autobiography of all the traumatic events rather than just specific events, as in CBT (Neuner et al., 2010). It aims to modify the associations of memories and representations of the traumatizing experiences in order to improve the survivors' condition. The promising outcome of NET was demonstrated in several studies on refugee population (Hensel-Dittmann et al., 2011; Robjant & Fazel, 2010; Stenmark, Catani, Neuner, Elbert, & Holen, 2013). It has been modified to suit the Southeast Asian context, by incorporating traditional values and beliefs, such as the Buddhist traditions for Khmer Rouge survivors (Strasser, Chhim, & Taing, 2014). The inclusion of spirituality in NET was found to increase its effectiveness (Agger, Igreja, Kiehle, & Polatin, 2012).

EMDR is another therapeutic modality that has gained ground in recent years. EMDR is a non-intrusive approach, which supports survivors in processing the elements such as thoughts, beliefs, emotions, images, and bodily sensations related to the trauma they have experienced. By connecting these elements related to the past, present, and future, EMDR allows survivors to safely review the experience in an adaptive way, which consequently changes the meaning and outcome of present triggers, and prepares schemas for future coping (Shapiro, 2001). It has shown to be effective in treating traumatized individuals (Lee & Cuijpers, 2013) and is one of World Health Organization's (2013) recommended treatments for PTSD. In Southeast Asia, EMDR is being promoted by international organizations as an effective and efficient treatment for humanitarian purposes (Mehrotra, 2014), with results at par with those of CBT in treating panic disorders (Horst et al., 2017). Its stabilization techniques are also found to be effective for Southeast Asian clients (Mattheß et al., 2018).

CHALLENGES IN MHPSS PROVISION IN SOUTHEAST ASIA

Southeast Asia has more than its fair share of disasters from various hazards. This puts an enormous strain on its disaster responders, putting them at risk of experiencing psychological distress. Fortunately, there are MHPSS services that may support them as they perform their tasks, but even the provision of these services often meet several obstacles. The following are just some of the challenges in providing MHPSS to Southeast Asian responders.

Human Resources

A major challenge worldwide, but particularly in the Region, is the lack of human resources. There are very few mental health professionals relative to the population in the Region. Data from 2006 reveal that the number of psychiatrists and psychologists, respectively, for every 100,000 people in Vietnam is 0.35 and 0.06, Thailand is 0.66 and 0.26, and the Philippines is 0.42 and 0.91 (Vuong, Ginneken, Morris, Ha, & Busse, 2011). This lack of access can be addressed in part by training more non-specialists to provide PFA and other non-specialized focused interventions. Along with increasing human resources, it is also possible to decrease the risk of developing serious psychological reactions. For example, lack of preparation also puts the responder at risk of developing serious psychological conditions. This is particularly a problem for non-traditional responders, who are not trained to tackle disaster stressors. In this respect, proper training and adequate preparation for responders would may reduce this risk.

Culture

As stated earlier, there are organizational factors particularly in male-dominated agencies that do not promote help-seeking behavior. Some emergency response organizations place a stigma on their members who are seeking for help (Haugen, McCrillis, Smid, & Nijdam, 2017). Furthermore, the collectivistic culture orientation of Southeast Asia may also hamper help-seeking (Chen et al., 2012). As collectivistic cultures value harmony, people espousing this culture orientation may hesitate to ask for help, as it may inconvenience others. MHPSS providers should make sure that they convey an open and non-judgmental stance; that they are available whenever the responder feels comfortable to seek for help.

Coordination

Another problem in providing MHPSS in the region is the lack of coordination among agencies. A major tenet of MHPSS is that it should not be a standalone service, but must be integrated with other programs and services. The lack of coordination between the different agencies and responders working in disaster settings makes the proper provision of MHPSS difficult. An offshoot of this issue is the proliferation of individuals and groups entering the disaster site and providing psychological intervention without coordinating with official agencies responsible for running post-disaster response. This should be avoided proper coordination can never be underemphasized in post-disaster settings. When MHPSS is delivered along with other services (e.g., medical) to disaster responders, it is easier to link the MHPSS recipient to other needed services. Furthermore, the disaster responder does not suffer from the fatigue of having to go through duplicating interventions when MHPSS is delivered in coordination with other services. Lack of coordination may add to the chaos, which ultimately does more harm than good.

Research-related Issues

Psychological interventions need to be backed by evidence of their effectiveness and efficiency. The need for evidence of an intervention's suitability to the Southeast Asian context is highlighted by the recognition of the difference between Western psychological support model and the local culture and worldviews (B. S. K. Kim, Atkinson, & Umemoto, 2001). However, there is still a dearth of rigorous research on Southeast Asia, more so, in responder groups. This is even exacerbated by the difficulty of conducting research in disaster settings. For example, most mental health research in disasters lack real pre-disaster measures (Norris & Elrod, 2006), which poses a challenge in isolating the effects of intervention from natural recovery. There is also the heavy reliance on self-report measures. Furthermore, many interventions conducted in the region are not well documented; hence, it is difficult to measure its effectiveness and to replicate them in other contexts.

CONCLUSION

This chapter attempted to summarize the MHPSS services in Southeast Asia. This chapter pointed out the types of responders working in the region and the challenges that they face. Though most responders will be able to resume normal functioning after deployment, a small number may develop serious psychological conditions such as PTSD, depression, and anxiety disorders. However, there are programs and activities that may mitigate these effects. Following the guidelines set by the IASC, a hierarchy of mental health interventions is presented; namely, the (1) provision of basic services; (2) strengthening of social networks; (3) provision of focused, non-specialized intervention; and (4) provision of specialized treatments. Although there are various innovations in this aspect, the conduct of MHPSS in the Region is still faced with a number of issues. Nevertheless, the recognition of these limitations provides a place where MHPSS providers and researchers may start in order to move forward.

REFERENCES

Adams, R. E., Boscarino, J. A., & Figley, C. R. (2006). Compassion fatigue and psychological distress among social workers: A validation study. *American Journal of Orthopsychiatry*, *76*(1), 103–108. https://doi.org/10.1037/0002-9432.76.1.103

Agger, I., Igreja, V., Kiehle, R., & Polatin, P. (2012). Testimony ceremonies in Asia: Integrating spirituality in testimonial therapy for torture survivors in India, Sri Lanka, Cambodia, and the Philippines. *Transcultural Psychiatry*, *49*(3–4), 568–589. https://doi.org/10.1177/1363461512447138

Alexander, D. A., & Klein, S. (2009). First responders after disasters: A review of stress reactions, at-risk, vulnerability, and resilience factors. *Prehospital and Disaster Medicine*. https://doi.org/10.1017/S1049023X00006610

Alvarez, J., & Hunt, M. (2005). Risk and resilience in canine search and rescue handlers after 9/11. *Journal of Traumatic Stress*, *18*(5), 497–505. https://doi.org/10.1002/jts.20058

Arnberg, F. K., Hultman, C. M., Michel, P.-O., & Lundin, T. (2012). Social support moderates posttraumatic stress and general distress after disaster. *Journal of Traumatic Stress*, *25*, 721–727. https://doi.org/10.1002/jts.21758

Below, R., & Wallemacq, P. (2018). *Annual disaster statistical review* 2017. Brussels: CRED.

Benedek, D. M., Fullerton, C. S., & Ursano, R. J. (2007). First responders: Mental health consequences of natural and human-made disasters for public health and public safety workers. *Annual Review of Public Health*, *28*, 55–68. https://doi.org/10.1146/annurev.publhealth.28.021406.144037

Benight, C. C., & Bandura, A. (2004). Social cognitive theory of posttraumatic recovery: The role of perceived self-efficacy. *Behaviour Research and Therapy*, *42*(10), 1129–1148. https://doi.org/10.1016/j.brat.2003.08.008

Berger, W., Coutinho, E. S. F., Figueira, I., Marques-Portella, C., Luz, M. P., Neylan, T. C., … Mendlowicz, M. V. (2012). Rescuers at risk: A systematic review and meta-regression analysis of the worldwide current prevalence and correlates of PTSD in rescue workers. *Social Psychiatry and Psychiatric Epidemiology*, *47*(6), 1001–1011. https://doi.org/10.1007/s00127-011-0408-2

Brewin, C. R., Andrews, B., & Valentine, J. D. (2000). Meta-analysis of risk factors for posttraumatic stress disorder in trauma-exposed adults. *Journal of Consulting and Clinical Psychology*, *68*(5), 748–766. https://doi.org/10.1037/0022-006X.68.5.748

Bromet, E. J., Hobbs, M. J., Clouston, S. A. P., Gonzalez, A., Kotov, R., & Luft, B. J. (2016). DSM-IV post-traumatic stress disorder among World Trade Center responders 11-13 years after the disaster of 11 September 2001 (9/11). *Psychological Medicine*, *46*(4), 771–783. https://doi.org/10.1017/S0033291715002184

Brooks, S. K., Dunn, R., Amlôt, R., Greenberg, N., & Rubin, G. J. (2016). Social and occupational factors associated with psychological distress and disorder among disaster responders: A systematic review. *BMC Psychology*, *4*(1), 18. https://doi.org/10.1186/s40359-016-0120-9

Bryant, R. A., Ekasawin, S., Chakrabhand, S., Suwanmitri, S., Duangchun, O., & Chantaluckwong, T. (2011). A randomized controlled effectiveness trial of cognitive behavior therapy for post-traumatic stress disorder in terrorist-affected people in Thailand. *World Psychiatry*, *10*(3), 205–209.

Bush, R. (2015). Muhammadiyah and disaster response: Innovation and change in humanitarian assistance. In C. Brassard, A. M. Howitt, & D. W. Giles (Eds.), *Natural disaster management in the Asia-Pacific: Policy and governance* (pp. 33–48). Tokyo: Springer Japan. https://doi.org/10.1007/978-4-431-55157-7_3

Chan, C. S., Tang, K. N. S., Hall, B. J., Yip, S. Y. T., & Maggay, M. (2016). Psychological sequelae of the 2013 Super Tyhpoon Haiyan among survivor-responders. *Psychiatry*, *79*(3), 282–296. https://doi.org/10.1080/00332747.2015.1129874

Chen, J. M., Kim, H. S., Mojaverian, T., & Morling, B. (2012). Culture and social support provision: Who gives what and why. *Personality and Social Psychology Bulletin*, *38*(1), 3–13. https://doi.org/10.1177/0146167211427309

Cieslak, R., Shoji, K., Douglas, A., Melville, E., Luszczynska, A., & Benight, C. C. (2014). A meta-analysis of the relationship between job burnout and secondary traumatic stress among workers with indirect exposure to trauma. *Psychological Services*, *11*(1), 75–86. https://doi.org/10.1037/a0033798

Creamer, T. L., & Liddle, B. J. (2005). Secondary traumatic stress among disaster mental health workers responding to the September 11 attacks. *Journal of Traumatic Stress*, *18*(1), 89–96. https://doi.org/10.1002/jts.20008

Cutrona, C. E., & Russell, D. W. (1990). Type of social support and specific stress: Toward a theory of optimal matching. In I. G. Sarason & G. R. Pierce (Eds.), *Social support: An interactional view* (pp. 319–366). Oxford: John Wiley.

De Soir, E. L. J. L. (2006). Psychosocial crisis intervention with military and emergency services personnel. In G. Reyes & G. A. Jacobs (Eds.), *Handbook of international disaster psychology* (Vol. 4, pp. 113–130). Westport, CT: Praeger.

de Terte, I., Stephens, C., & Huddleston, L. (2014). The development of a three part model of psychological resilience. *Stress and Health*, *30*(5), 416–424. https://doi.org/10.1002/smi.2625

Deelstra, J. T., Peeters, M. C. W., Schaufeli, W. B., Stroebe, W., Zijlstra, F. R. H., & van Doornen, L. P. (2003). Receiving instrumental support at work: When help is not welcome. *Journal of Applied Psychology*, *88*(2), 324–331. https://doi.org/10.1037/0021-9010.88.2.324

Dunn, R., Brooks, S. K., & Greenberg, N. (2015). Psychological impact of traumatic events: guidance for trauma-exposed organizations. *Occupational Health at Work*, *12*(1), 17–21.

Flores, J., Melgar, I., Guiang, R., Jopson, Q., Docena, P., Fornillos, J., & Pan, M. (2014, August). Developing Resilience among college students in Tacloban. In M. R. Hechanova (Chair),

Enabling resilience and recovery: Interventions and lessons. Symposium conducted in the 51st Annual Convention of the Psychological Association of the Philippines, Cagayan de Oro, Philippines.

Franz, M. R., Wolf, E. J., MacDonald, H. Z., Marx, B. P., Proctor, S. P., & Vasterling, J. J. (2013). Relationships among predeployment risk factors, warzone-threat appraisal, and postdeployment PTSD symptoms. *Journal of Traumatic Stress*, *26*(4). https://doi.org/10.1002/jts.21827

Gabert-Quillen, C. A., Irish, L. A., Sledjeski, E., Fallon, W., Spoonster, E., & Delahanty, D. L. (2012). The impact of social support on the relationship between trauma history and PTSD symptoms in motor vehicle accident victims. *International Journal of Stress Management*, *19*(1), 69–79. https://doi.org/10.1037/a0026488

Guha-Sapir, D., Hoyois, P., & Below, R. (2016). *Annual disaster statistical review 2015: The numbers and trends*. Brussels: CRED. Retrieved from http://www.cred.be/sites/default/files/ADSR_2015.pdf

Guilaran, J. B. (2013). What you "don't see" can't hurt you: The role of attachment avoidance in compassion fatigue among humanitarian relief workers in the Philippines. *Philippine Journal of Social Sciences and Humanities*, *18*(1), 11–21.

Guilaran, J., de Terte, I., Kaniasty, K., & Stephens, C. (2018). Psychological outcomes in disaster responders: A systematic review and meta-analysis on the effect of social support. *International Journal of Disaster Risk Science*, *9*(3), 344–358. https://doi.org/10.1007/s13753-018-0184-7

Halbesleben, J. R. B. (2006). Sources of social support and burnout: A meta-analytic test of the conservation of resources model. *Journal of Applied Psychology*, *91*(5), 1134–1145. https://doi.org/10.1037/0021-9010.91.5.1134

Haslam, C., & Mallon, K. (2003). A preliminary investigation of post-traumatic stress symptoms among firefighters. *Work & Stress*, *17*(3), 277–285. https://doi.org/10.1080/02678370310001625649

Haugen, P. T., Evces, M., & Weiss, D. S. (2012). Treating posttraumatic stress disorder in first responders: A systematic review. *Clinical Psychology Review*, *32*(5), 370–380. https://doi.org/10.1016/j.cpr.2012.04.001

Haugen, P. T., McCrillis, A. M., Smid, G. E., & Nijdam, M. J. (2017). Mental health stigma and barriers to mental health care for first responders: A systematic review and meta-analysis. *Journal of Psychiatric Research*, *94*, 218–229. https://doi.org/10.1016/j.jpsychires.2017.08.001

Hechanova, R. M., Ramos, P. A. P., & Waelde, L. C. (2015). Group-based mindfulness- informed psychological first aid after Typhoon Haiyan. *Disaster Prevention and Management*, 24(5): 610–618.

Hechanova, R., & Waelde, L. (2017). The influence of culture on disaster mental health and psychosocial support interventions in Southeast Asia. *Mental Health, Religion & Culture*, *20*(1), 31–44. https://doi.org/10.1080/13674676.2017.1322048

Hechanova, M. R. M., Waelde, L., Docena, P., Alampay, L. P., Alianan, A. S., Flores, M. J. B., Ramos, P. A. P., & Melgar, M. I. E. (2016). The development and initial evaluation of *Katatagan*: A resilience intervention for Filipino disaster survivors. *Philippine Journal of Psychology*, *48*(2), 105–131.

Hechanova, M. R. M., Waelde, L. C., & Ramos, P. A. P. (2016). Evaluation of a group-based resilience intervention for Typhoon Haiyan survivors. *Journal of Pacific Rim Psychology*, *10*, e12. https://doi.org/10.1017/prp.2016.9

Hensel-Dittmann, D., Schauer, M., Ruf, M., Catani, C., Odenwald, M., Elbert, T., & Neuner, F. (2011). Treatment of traumatized victims of war and torture: A randomized controlled comparison of narrative exposure therapy and stress inoculation training. *Psychotherapy and Psychosomatics*, *80*(6), 345–352. https://doi.org/10.1159/000327253

Hinton, D. E., Chhean, D., Pich, V., Safren, S. A., Hofmann, S. G., & Pollack, M. H. (2005). A randomized controlled trial of cognitive-behavior therapy for Cambodian refugees with treatment-resistant PTSD and panic attacks: A cross-over design. *Journal of Traumatic Stress*, *18*(6), 617–629. https://doi.org/10.1002/jts.20070

Hinton, D. E., Pham, T., Tran, M., Safren, S. A., Otto, M. W., & Pollack, M. H. (2004). CBT for Vietnamese refugees with treatment-resistant PTSD and panic attacks: a pilot study. *Journal of Traumatic Stress*, *17*(5), 429–433. https://doi.org/10.1023/B:JOTS.0000048956.03529.fa

Hobfoll, S. E. (1989). Conservation of resources. A new attempt at conceptualizing stress. *The American Psychologist*, *44*(3), 513–524. https://doi.org/10.1037/0003-066X.44.3.513

Hobfoll, S. E., Freedy, J., Lane, C., & Geller, P. (1990). Conservation of social resources: Social support resource theory. *Journal of Personal and Social Relationships*, *7*, 465–478. https://doi.org/0803973233

Hobfoll, S. E., & Stokes, J. P. (1988). The process and mechanics of social support. In S. Duck, D. F. Hay, S. E. Hobfoll, W. Ickes, & B. M. Montgomery (Eds.), *Handbook of personal relationships: theory, research and interventions* (pp. 497–517). Oxford: John Wiley & Sons.

Hobfoll, S. E., Watson, P., Bell, C. C., Bryant, R. A., Brymer, M. J., Friedman, M. J., … Ursano, R. J. (2007). Five essential elements of immediate and mid-term mass trauma intervention: empirical evidence. *Psychiatry*, *70*(4), 283–315; discussion 316-369. https://doi.org/10.1521/psyc.2007.70.4.283

Horst, F., Den Oudsten, B., Zijlstra, W., de Jongh, A., Lobbestael, J., & De Vries, J. (2017). Cognitive behavioral therapy vs. eye movement desensitization and reprocessing for treating panic disorder: A randomized controlled trial. *Frontiers in Psychology*, *8*, 1409. https://doi.org/10.3389/fpsyg.2017.01409

Inter-agency Standing Committee (IASC). (2007). *IASC guidelines on mental health and psychosocial support in emergency settings*. Geneva.

Iserson, K. V., Heine, C. E., Larkin, G. L., Moskop, J. C., Baruch, J., & Aswegan, A. L. (2008). Fight or flight: The ethics of emergency physician disaster response. *Annals of Emergency Medicine*, *51*(4), 345–353. https://doi.org/10.1016/j.annemergmed.2007.07.024

Kaniasty, K., & Norris, F. H. (1995). Mobilization and Deterioration of Social Support Following Natural Disasters. *Current Directions in Psychological Science*, *4*(3), 94–98. https://doi.org/10.1111/1467-8721.ep10772341

Kaniasty, K., & Norris, F. H. (2000). Help-seeking comfort and receiving social support: The role of ethnicity and context of need. *American Journal of Community Psychology*, *28*(4), 545–581. https://doi.org/10.1023/A:1005192616058

Kaniasty, K., & Norris, F. H. (2004). Social support in the aftermath of disasters, catastrophes, and acts of terrorism: altruistic, overwhelmed, uncertain, antagonistic, and patriotic communities. In R. J. Ursano, A. E. Norwood, & C. S. Fullerton (Eds.), *Bioterrorism: Psychological and public health interventions* (pp. 200–229). Cambridge: Cambridge University Press.

Kaniasty, K., & Norris, F. H. (2009). Distinctions that matter: Received social support, perceived social support, and social embeddedness after disasters. In Y. Neria, S. Galea, & F. H. Norris (Eds.), *Mental Health and Disasters* (pp. 175–200). New York, NY: Cambridge University Press.

Kim, B. S. K., Atkinson, D. R., & Umemoto, D. (2001). Asian cultural values and the counseling process: Current knowledge and directions for future research. *The Counseling Psychologist*, *29*(4), 570–603. https://doi.org/10.1177/0011000001294006

Kim, H. S., Sherman, D. K., & Taylor, S. E. (2008). Culture and social support. *The American Psychologist*, *63*(6), 518–526. https://doi.org/10.1037/0003-066X

Kleim, B., & Westphal, M. (2011). Mental health in first responders: A review and recommendation for prevention and intervention strategies. *Traumatology*, *17*(4), 17–24. https://doi.org/10.1177/1534765611429079

Ladrido-Ignacio, L. (2011). Responders to disasters. In *Ginhawa: Wellbeing in the Aftermath of Disasters*. Pasig City: Philippine Psychiatrists Association, Inc.

Lai, C-H, Chib, A., & Ling, R. (2018). Digital disparities and vulnerability: Mobile phone use, information behaviour, and disaster preparedness in Southeast Asia. *Disasters*, *42*(4), 734–760.

Landoy, B. V. N., Hechanova, M. R. M., Ramos, P. A. P., & Kintanar, N. S. M. (2015). The application and adaptation of psychological first aid: The Filipino psychologists' experience after Typhoon Haiyan. *Philippine Journal of Psychology*, *48*(2), 81–104.

Lee, C. W., & Cuijpers, P. (2013). A meta-analysis of the contribution of eye movements in processing emotional memories. *Journal of Behavior Therapy and Experimental Psychiatry*, *44*(2), 231–239. https://doi.org/10.1016/j.jbtep.2012.11.001

Lindgaard, C. V., Iglebaek, T., & Jensen, T. K. (2009). Changes in family functioning in the aftermath of a natural disaster: The 2004 tsunami in Southeast Asia. *Journal of Loss and Trauma*, *14*(2), 101–116. https://doi.org/10.1080/15325020802537138

Ludick, M., & Figley, C. R. (2017). Toward a mechanism for secondary trauma induction and reduction: Reimagining a theory of secondary traumatic stress. *Traumatology*, 23(1), 112–123.

Maslach, C., Schaufeli, W. B., & Leiter, M. P. (2001). Job burnout. *Annual Review of Psychology, 52*, 397–422. https://doi.org/10.1146/annurev.psych.52.1.397

Mattheß, C., Farrell, D., Mattheß, M., Bumke, P., Sodemann, U., & Mattheß, H. (2018). The therapeutic value of trauma stabilisation in the treatment of Post-traumatic stress disorder – A Southeast Asian Study. *Asian Journal of Psychiatry, 41*, 45–49. https://doi.org/10.1016/j.ajp.2018.09.010

McFarlane, A. C., & Norris, F. H. (2006). Definitions and concepts in disaster research. In F. H. Norris, S. Galea, M. J. Friedman, & P. J. Watson (Eds.), *Methods for disaster mental health research* (pp. 3–19). New York, NY: The Guilford Press.

Mehrotra, S. (2014). Humanitarian projects and growth of EMDR therapy in Asia. *Journal of EMDR Practice and Research, 8*, 252–259. https://doi.org/10.1891/1933-3196.8.4.252

National Institute for Health and Clinical. (2005). *Post traumatic stress disorder: The management of PTSD in adults and children in primary and secondary care*. London: Gaskell and the British Psychological Society.

Naturale, A., & Pulido, M. L. (2012). Helping the helpers: Ameliorating secondary traumatic stress in disaster workers. In J. L. Framingham & M. L. Teasley (Eds.), *Behavioral health response to disasters* (pp. 189–208). Boca Raton, FL: CRC Press.

Neuner, F., Kurreck, S., Ruf, M., Odenwald, M., Elbert, T., & Schauer, M. (2010). Can asylum-seekers with posttraumatic stress disorder be successfully treated? A randomized controlled pilot study. *Cognitive Behaviour Therapy, 39*(2), 81–91. https://doi.org/10.1080/16506070903121042

Nguyen, Q. (2016). Do natural disasters open a window of opportunity for corruption? *The Journal of Development Studies* (Vol. 53). https://doi.org/10.1080/00220388.2016.1187724

Norris, F. H., & Elrod, C. L. (2006). Psychosocial consequences of disaster: A review of past research. In F. H. Norris, S. Galea, M. J. Friedman, & P. J. Watson (Eds.), *Methods for disaster mental health research* (pp. 20–42). New York: The Guilford Press.

Norris, F. H., Friedman, M. J., Watson, P. J., Byrne, C. M., Diaz, E., & Kaniasty, K. (2002). 60,000 disaster victims speak: Part I. An empirical review of the empirical literature, 1981–2001. *Psychiatry, 65*(3), 207–239. https://doi.org/10.1521/psyc.65.3.207.20173

Norris, F. H., & Kaniasty, K. (1996). Received and perceived social support in times of stress: A test of the social support deterioration deterrence model. *Journal of Personality and Social Psychology, 71*(3), 498–511.

North, C. S., & Pfefferbaum, B. (2013). Mental health response to community disasters. *Clinical Review & Education, 310*(5), 507–518. https://doi.org/10.1001/jama.2013.107799

North, C. S., Tivis, L., McMillen, J. C., Pfefferbaum, B., Cox, J., Spitznagel, E. L., … Smith, E. M. (2002). Coping, functioning, and adjustment of rescue workers after the Oklahoma City bombing. *Journal of Traumatic Stress, 15*(3), 171–175. https://doi.org/10.1023/A:1015286909111

Otto, M. W., & Hinton, D. E. (2006). Modifying exposure-based CBT for Cambodian refugees with posttraumatic stress disorder. *Cognitive and Behavioral Practice, 13*(4), 261–270. https://doi.org/10.1016/j.cbpra.2006.04.007

Ozer, E. J., Best, S. R., Lipsey, T. L., & Weiss, D. S. (2003). Predictors of posttraumatic stress disorder and symptoms in adults: A meta-analysis. *Psychological Bulletin, 129*(1), 52–73. https://doi.org/10.1037/0033-2909.129.1.52

Palic, S., & Elklit, A. (2011). Psychosocial treatment of posttraumatic stress disorder in adult refugees: A systematic review of prospective treatment outcome studies and a critique. *Journal of Affective Disorders, 131*(1), 8–23. https://doi.org/10.1016/j.jad.2010.07.005

Palm, K. M., Polusny, M. A., & Follette, V. M. (2004). Vicarious traumatization: Potential hazards and interventions for disaster and trauma workers. *Prehospital and Disaster Medicine, 19*(1), 73–78.

Panyayong, B., & Pengjuntr, W. (2006). Mental health and psychosocial aspects of disaster preparedness in Thailand. *International Review of Psychiatry, 18*(6), 607–614. https://doi.org/10.1080/09540260601038514

Paton, D. (1996). Training disaster workers: Promoting wellbeing and operational effectiveness. *Disaster Prevention and Management, 5*(5), 11–18. https://doi.org/10.1108/09653569610131735

Pekevski, J. (2013). First responders and psychological first aid. *Journal of Emergency Management, 11*(1), 39–48. https://doi.org/10.5055/jem.2013.0126

Pietrzak, R. H., Feder, A., Singh, R., Schechter, C. B., Bromet, E. J., Katz, C. L., … Southwick, S. M. (2014). Trajectories of PTSD risk and resilience in World Trade Center responders: An 8-year prospective cohort study. *Psychological Medicine*, *44*(1), 205–219. https://doi.org/10.1017/S0033291713000597

Pityaratstian, N., Piyasil, V., Ketumarn, P., Sitdhiraksa, N., Ularntinon, S., & Pariwatcharakul, P. (2015). Randomized controlled trial of group cognitive behavioural therapy for post-traumatic stress disorder in children and adolescents exposed to tsunami in Thailand. *Behavioural and Cognitive Psychotherapy*, *43*(5), 549–561. https://doi.org/10.1017/S1352465813001197

Prati, G., & Pietrantoni, L. (2010a). An application of the social support deterioration deterrence model to rescue workers. *Journal of Community Psychology*, *38*(7), 901–917. https://doi.org/10.1002/jcop.20404

Prati, G., & Pietrantoni, L. (2010b). The relation of perceived and received social support to mental health among first responders: A meta-analytic review. *Journal of Community Psychology*, *38*(3), 403–417. https://doi.org/10.1002/jcop.20371

Prati, G., Pietrantoni, L., & Cicognani, E. (2010). Self-efficacy moderates the relationship between stress appraisal and quality of life among rescue workers. *Anxiety, Stress & Coping: An International Journal*, *23*(4), 463–470. https://doi.org/10.1080/10615800903431699

Puig, M. E., & Glynn, J. B. (2004). Disaster responders: A cross-cultural approach to recovery and relief work. *Journal of Social Service Research*, *30*(2), 55–66. https://doi.org/10.1300/J079v30n02_04

Rini, C., & Dunkel Schetter, C. (2010). The effectiveness of social support attempts in intimate relationships. In K. Sullivan & J. Davila (Eds.), *Support processes in intimate relationships* (pp. 26–67). New York: Oxford University Press.

Rini, C., Dunkel Schetter, C., Hobel, C. J., Glynn, L. M., & Sandman, C. A. (2006). Effective social support: Antecedents and consequences of partner support during pregnancy. *Personal Relationships*, *13*, 207–229.

Robjant, K., & Fazel, M. (2010). The emerging evidence for narrative exposure therapy: A review. *Clinical Psychology Review*, *30*(8), 1030–1039. https://doi.org/10.1016/j.cpr.2010.07.004

Ruzek, J. I., Brymer, M. J., Jacobs, A. K., Layne, C. M., Vernberg, E. M., & Watson, P. J. (2007). Psychological first aid. *Journal of Mental Health Counseling*, *29*(1), 17–49. https://doi.org/10.17744/mehc.29.1.5racqxjuefabgwp

Shalev, A. Y. (2000). Biological responses to disasters. *Psychiatric Quarterly*, *71*(3), 277–288. https://doi.org/10.1023/A:1004686211979

Shapiro, F. (2001). *Eye movement desensitization and reprocessing (EMDR): Basic principles, protocols, and procedures*. New York: Guilford Press.

Stenmark, H., Catani, C., Neuner, F., Elbert, T., & Holen, A. (2013). Treating PTSD in refugees and asylum seekers within the general health care system. A randomized controlled multicenter study. *Behaviour Research and Therapy*, *51*(10), 641–647. https://doi.org/10.1016/j.brat.2013.07.002

Strasser, J., Chhim, S., & Taing, S. (2014). Narrative exposure therapy (NET): Culturally sensitive trauma treatment for Khmer Rouge survivors. *The Journal of the International Association of Buddhist Universities*, *7*(1), 100–104.

Tam, C. W. C., Pang, E. P. F., Lam, L. C. W., & Chiu, H. F. K. (2004). Severe acute respiratory syndrome (SARS) in Hong Kong in 2003: Stress and psychological impact among frontline healthcare workers. *Psychological Medicine*, *34*(7), 1197–1204. https://doi.org/10.1017/S0033291704002247

Taylor, S. E., Sherman, D. K., Kim, H. S., Jarcho, J., Takagi, K., & Dunagan, M. S. (2004). Culture and social support: Who seeks it and why? *Journal of Personality and Social Psychology*, *87*(3), 354–362. https://doi.org/10.1037/0022-3514.87.3.354

Tedeschi, R. G., & Calhoun, L. G. (2004). Posttraumatic growth: Conceptual foundations and empirical evidence. *Psychological Inquiry*, *15*(1), 1–18. https://doi.org/10.1207/s15327965pli1501

Thormar, S. B., Gersons, B. P. R., Juen, B., Djakababa, M. N., Karlsson, T., & Olff, M. (2013). Organizational factors and mental health in community volunteers. The role of exposure, preparation, training, tasks assigned, and support. *Anxiety, Stress, and Coping*, *26*(6), 624–642. https://doi.org/10.1080/10615806.2012.743021

Udomratn, P. (2008). Mental health and the psychosocial consequences of natural disasters in Asia. *International Review of Psychiatry*, *20*(5), 441–444. https://doi.org/10.1080/09540260802397487

Vernberg, E. M., Steinberg, A. M., Jacobs, A. K., Brymer, M. J., Watson, P. J., Osofsky, J. D., … Ruzek, J. I. (2008). Innovations in disaster mental health: Psychological first aid. *Professional Psychology: Research and Practice*, *39*(4), 381–388. https://doi.org/10.1037/a0012663

Villaflor, P. I. A. T. M., Villaret, N. P. B., & Hechanova, M. R. M. (2019). The role of Filipino masculine ideology on the adaptive coping, psychological well-being and vicarious trauma of first responders. *International Journal of Culture and Mental Health*, 1–10. https://doi.org/10.1080/17542863.2018.1561736

Vuong, D. A., Ginneken, E. Van, Morris, J., Ha, S. T., & Busse, R. (2011). Mental health in Vietnam: Burden of disease and availability of services. *Asian Journal of Psychiatry*, *4*(1), 65–70. https://doi.org/10.1016/j.ajp.2011.01.005

Waelde, L. C. (2014). *Inner resources for stress* (3rd ed.). Palo Alto, CA: Treatment Manual and Materials, Inner Resources Center, Palo Alto University.

Waelde, L. C., Uddo, M., Marquette, R., Ropelato, M., Freightman, S., Pardo, A., & Salazar, J. (2008). A pilot study of meditation for mental health workers following Hurricane Katrina. *Journal of Traumatic Stress*, *12*(5), 497–500.

Welle, T. & Birkmann, J. (2015). The world risk index: An approach to assess risk and vulnerability on a global scale. *Journal of Extreme Events*, *2*(1), 1550003. https://doi.org/10.1142/S2345737615500037

Woo, C. K., Cheng, Y. S., Ho, T., & Horowitz, I. (2015). Public trust and corruption perception: disaster relief. *Applied Economics*, *47*(46), 4967–4981. https://doi.org/10.1080/00036846.2015.1039703

World Health Organization. (2013). *Guidelines for the managements of conditions specifically related to stress*. Geneva: Author.

Wyatt, Z., Hoban, E., & Macfarlane, S. (2018). Trauma-Informed Education Practice in Cambodia. *International Journal of Asian Social Science*, *8*(2), 62–76.

CHAPTER 9

POSTTRAUMATIC GROWTH AND RESILIENCE IN SOUTHEAST ASIA

Grant J. Rich and Skultip (Jill) Sirikantraporn

ABSTRACT

After decades of focus on disaster, crisis, and trauma itself, in recent years more attention has been devoted to the study of human strengths and resilience, as reflected in the rise of positive psychology and strengths-based social work. In particular, psychological growth after trauma has been increasingly studied, and one of the official terms referring to the phenomenon is posttraumatic growth (PTG). The PTG literature reflects work on positive psychology, trauma recovery, and resilience. The main components associated with PTG are new possibilities, interpersonal growth, personal growth, appreciation for life, and spiritual change (Calhoun & Tedeschi, 2014). These domains have been tested and measured with a scale, the Posttraumatic Growth Inventory. While PTG and related concepts such as resilience have been studied in various populations, they have not yet been investigated extensively in Southeast Asia (SEA) populations. This chapter explores the psychological examination of resilience and PTG in the SEA context, with some discussion of the background of both positive psychological concepts and PTG research cross-culturally, and their application to the SEA region specifically. Brief relevant trauma history of the region, such as human-made and natural hazards impacting the region's individuals and communities, and similarities and differences in the results of these traumas will be described. Implications for broader international work as well as cultural and clinical implications also will be discussed in this chapter.

Keywords: Trauma; resilience; posttraumatic growth; Southeast Asia; PTSD; spirituality

Resistance, Resilience, and Recovery from Disasters: Perspectives from Southeast Asia
Community, Environment and Disaster Risk Management, Volume 21, 143–158

ISSN: 2040-7262/doi:10.1108/S2040-726220200000021009

Historically, psychology, generally speaking, has focused on human suffering. For instance, much research has been focused on those with psychological disorders, on those in emotional pain, and on how best to assist them. Such organized efforts have yielded many great successes, and evidence-based improvements beyond reliance on anecdotal evidence or worse, active discrimination and prejudice toward those with mental illness. Yet, over the past 20 years or so, it is been recognized that careful examination of human strengths and well-being and those who are thriving may add critical, even essential, insights into our understanding of the human condition (Rich & Sirikantraporn, 2018). This trend in positive psychology and related work, led by Martin E.P. Seligman, Mihaly Csikszentmihalyi, Chris Peterson, Shane Lopez, and others, has resulted in a number of advancements in psychology. Notably, by emphasizing human strengths and virtues, it has provided a critical corrective to what has been termed as the business-as-usual or doom and gloom psychology that has been the principle focus of the majority the psychology discipline since the nineteenth century, thus linking the previous history of deficit-focused psychology to current trends which are more strengths-focused (e.g., Peterson, 2006; Seligman, 2011).

Positive psychologists argue for what they view as a methodologically rigorous, evidence-based, scientific approach and actively aim to avoid a naïve or overly optimistic view of humanity (Rich, 2016, 2017; Seligman, 2002; Seligman & Csikszentmihalyi, 2000). Topics considered by positive psychologists are numerous and include positive emotions, positive traits, and positive institutions, including such foci as happiness, life satisfaction, wisdom, creativity, religion/spirituality, love, subjective well-being, positive coping, and resilience. It should be noted that scholarship in strength-based social work (e.g., Saleeby, 2008) also relates to positive psychology, as do the recently emerging subdisciplines in anthropology such as positive anthropology and anthropology of well-being (e.g., Mathews & Izquierdo, 2009).

Despite many accomplishments, positive psychology research has remained focused predominantly in the United States and Western Europe with some exceptions, such as work in China, India, and Japan (e.g., Lomas, 2018; Mathews & Izquierdo, 2009; Oishi, 2010). The international landscape of psychology has been changing over the past several decades and psychology is emerging from a discipline dominated by the United States to a much more international discipline. Such internationalization is a welcome development, given research demonstrating both the paucity of human universals (Brown, 1991) and the perils of research that only examines Western samples (Henrich, Heine, & Norenzayan, 2010). As evidence for the global expansion of psychology, one may examine estimates from Zoma and Gielen (2015), who found that about 76%–78% of the world's approximately one million psychologists now live outside the United States. Nevertheless, most textbooks in the field continue to rely almost exclusively on research conducted in North America and Europe (Takooshian, Gielen, Plous, Rich, & Velayo, 2016) despite recent trends and efforts to internationalize the teaching of psychology (e.g., Leong, Pickren, Leach, & Marsella, 2012; Rich, Gielen, & Takooshian, 2017).

Our chapter aims to bridge between these two rapidly expanding disciplines – positive psychology and international psychology – by focusing on the critical issue of human strengths, in particular posttraumatic growth (PTG) and resilience, in cross-cultural and global perspective, in particular, in Southeast Asia (SEA). Whereas it may be argued that there is a virtual cottage industry of books and articles on trauma, anxiety, depression, and posttraumatic stress disorder (PTSD), the numbers of books and articles on resilience and PTG is much smaller, and indeed we could only locate a handful of books if we specifically aim to examine texts that focus on PTG and resilience in cross-cultural context.

Given this background and context, it is useful to briefly describe several books that are relevant to the present purpose, in that they represent the few books on postraumatic growth and resilience with a cultural or international emphasis, though none of these books fully focuses on Asia or SEA. The books include *Youth Resilience and Culture: Commonalities and Complexities (Cross-cultural Advancements in Positive Psychology)* which examines nations such as Canada, Zimbabwe, New Zealand, China, and Colombia, and has a focus on youth only (Theron, Liebenberg, & Ungar, 2015). In *Posttraumatic Growth and Culturally Competent Practice: Lessons Learned from Around the Globe*, the two social work faculty authors focus on very different nations than those in SEA (Weiss & Berger, 2010). For example, their text examines PTG in Spain, the Netherlands, Israel, Japan, China, Germany, and Australia; only two of the nations (Japan and China) are clearly Asian, though some place Australia in that orbit as well. A seminal text, the book focuses on strengths-based social work perspectives that relate to similar approaches related to positive psychology. Notably, these authors found that across the cultures they examined, PTG relates to coping strategies, cognitive processing, social support, and religiosity/spirituality, albeit with some variations. For example, there are cultural nuances, such as receptivity to psychotherapy, related to varied cultural beliefs around the globe on stigma and mental illness. Though about 15 years old, *Trauma and Dissociation in a Cross-cultural Perspective* addresses some related content concerning trauma, though it focuses specifically on dissociation, and their research in large part historically came prior the contemporary movements in positive psychology and strengths-based work on PTG and resilience (Rhoades & Sar, 2005). Additionally, its focus is on nations typically very different from much of SEA (e.g., Argentina, China, Ireland, Australia, France, and Japan). Finally, Rich and Sirikantraporn's (2018) *Human Strengths and Resilience: Developmental, Cross-cultural and International Perspectives* focuses on many developing nations, with some in South Asia or SEA (including chapters on Cambodia and India), though most of the chapters are outside of Asia (e.g., Lebanon, Guatemala, Taiwan, South Africa, and Haiti). In addition to describing mixed methods work in these nations documenting PTG internationally, the scholars noted the role of culturally informed international humanitarian aid work in assisting clients and communities in the process of healing from traumatic events, utilizing their own personal and community resources and strengths along with guidance from experienced trauma psychologists. Understanding the local conditions and cultures and developing train the trainer programs utilizing local leaders are among the successful techniques described in the book (Rich & Sirikantraporn, 2018).

Several other valuable books examine PTG but do not have an international orientation. There are a number of books that explore PTG though without an explicit culture or international focus. Calhoun and Tedeschi's *Handbook of Posttraumatic Growth: Research and Practice* (2006) has no international focus, and Calhoun and Tedeschi's more recent *Posttraumatic Growth in Clinical Practice* (2013), includes only one chapter on "multicultural" issues. However, Calhoun and Tedeschi (2013) did describe other elements relevant to PTG in some depth; for instance, they wrote in detail about emotional regulation, constructive self-disclosure and redevelopment of relationships, creating a narrative with PTG domains, the role of expert companions, and existential, religious, and spiritual growth. *Trauma, Recovery, and Growth: Positive Psychological Perspectives on Posttraumatic Stress* takes a strengths-based approach, which is in keeping with positive psychology, but it includes only one chapter that can be arguably considered international (one on immigrants) (Joseph & Linley, 2008). *Promoting Capabilities to Manage Posttraumatic Stress: Perspectives on Resilience*, has just one international chapter (on South Africa), and does not adopt an explicitly positive psychology approach (Paton, Violanti, & Smith, 2002). *The Body Keeps Score: Brain, Mind, and Body in the Healing of Trauma* describes influential work, though is mostly focused on the United States, and does not explicitly employ or clearly utilize a strengths-based positive psychology approach (Van der Kolk, 2015). This work is noteworthy for its emphasis on how non-psychological interventions, such as judicious use of evidence-based complementary and alternative health techniques such as yoga and massage therapy may promote PTG and healing after trauma. Van der Kolk does not advocate abandoning use of psychotherapy; instead, he notes the positive effects of supplementing psychotherapy with additional appropriate integrative modalities, or utilizing such modalities when psychotherapy is not available or not advised.

Additionally, although this chapter focuses more narrowly in scope on PTG as a response to trauma, it is helpful to note that there is a genre of books that essentially focus on trauma in specific nations. These books often are related to a particular event, such as a disaster or violent conflict, but such books do not focus more generally around the globe in multiple nations. At present, there is no single volume book clearly focused on PTG and resilience in SEA. Notably, though these books focused on specific nations may be helpful in many ways, they often do not adopt a positive psychology strengths-based approach focused on PTG and resilience, rather they focus mostly on PTSD, and on other anxiety and depression issues. Furthermore, books focused upon non-Asian and non-Southeast Asian populations are likely to miss particular cultural nuances in trauma experiences and resilient responses. For example, the ways in which PTG manifests itself in a Southeast Asian nation may differ from the ways in which PTG may manifest itself in other culture. Specific examples of such cultural differences may be the role that Khmer beliefs in religion and karma, including both Buddhist beliefs and indigenous beliefs in spirits, impact resilience in different ways than beliefs in Catholicism or vodou do for Haitians. Thus, for instance, research exploring Haiti after the 2010 earthquake (Farmer, 2011), may not be as relevant as explanations for Cambodians' resilient coping addressing the impact of the era post Pol Pot (Van Schaack, Reicherter, & Chhang, 2011).

Thus, briefly put, the present chapter aims to meet an important need by bridging the rapidly expanding disciplines of positive psychology (and strengths-based social work/positive anthropology) with a focus on PTG and resilience as responses to trauma, with a much needed focus on SEA, in nations that clearly have underserved, understudied populations with a significant prevalence of trauma. The cultural differences in the region, with respect to language, religion, economics, customs, and beliefs about mental health offer further justification for why research with Southeast Asian populations is especially needed (e.g., Rich, Jaafar, & Barron, in press).

TRAUMA SOURCES

Though this chapter focuses on resilience and PTG, it is helpful to note as context the types of traumatic events that occur in the region. One motivator for a region-specific review is that the burden of disaster and conflict in SEA and the relative lack of infrastructure to respond leads to significant gaps in services and in appropriate response (e.g., Rich et al. in press; Tang, 2006). The full gamut of traumatic experiences presents itself in SEA, from natural hazards to human-made ones. For instance, in terms of natural hazards, there are numerous floods, earthquakes, heatwaves, and similar such events (e.g., Rich et al., in press; Tang, 2006). Notably, the 2004 earthquake and tsunami resulted in approximately 225,000 to 320,000 deaths, across 14 nations, including Thailand, Myanmar, and Malaysia (Hyndman, 2011; Thavichachart et al., 2009). Human-made disasters include political violence, war, terrorism, pollution, and traffic accidents. Additionally, smaller scale traumatic events include crime (physical and sexual assault), family violence, injury and accidents, life-threatening illnesses, fires, explosions, and so on. Other disasters, crises, and traumas resulting from corruption, including organized crime such as prostitution and policies leading to poverty and famine, may reflect a combination of individual and group factors (Kristof & WuDunn, 2009). Furthermore, in addition to direct traumatic experiences, many scholars have investigated the vicarious traumas of witnessing such events happen to others (Rich & Sirikantraporn, 2018) and generational trauma (Ferrara, 2018). A tragic example of a human-made disaster and the generational trauma that follows is the era of the Khmer Rouge regime and its dictator Pol Pot. The Khmer Rouge time was at its height between 1975 and 1979, killing approximately 2 million people, which represented roughly a quarter of the entire Cambodian population at that time, from execution, disease, starvation, and overwork (Chandler, 2007; Hinton, 2004; Kierman, 2008). Today, 40 years later, Cambodia continues to be plagued by the impact of that trauma, which impacts so many in that generation and in the generations that followed, and by cycles of violence and victimization due to continued political and/or social instability, high rates of poverty, and often very limited access to resources (Jarl, Cantor-Graae, Chak,

Sunbaunat, & Larsson, 2015; Sirikantraporn, Rich, & Jafari, 2018). A survey of approximately 3,000 adults from nine of the 25 provinces in Cambodia revealed that about 31% of women and 18.4% of men suffered from probable anxiety disorders. Furthermore, the survey found that 19.7% of women and 10.2% of men suffer from likely depressive disorders. This survey also revealed that among those who experienced the Civil War, 92.3% experienced one or more traumatic events (though the mean was seven events), and that among younger participants who did not experience the Civil War, 70.2% experienced at least one traumatic event (Royal University of Phnom Penh (RUPP), 2012). Such data indicate that even four decades post the peak of Pol Pot's Khmer Rouge, nearly three-fourths of younger Cambodians experience trauma, as well as frequent poverty and the impacts of displacement.

After large-scale natural and human-made disasters such as those noted above, it is common for behavioral health conditions of individuals and communities, including PTSD, to increase. Thus, an understanding of mental health work and how it relates to trauma and to PTG is critical. As an example, given its dramatic scope and scale, and its enduring impact in SEA, the following section will focus on a brief literature review of the mental health work related to trauma and resilience in Thailand following the Indian Ocean Tsunami in 2004 that killed about 320,000 human beings in total and over 5,000 in Thailand (Thavichachart et al., 2009). Because timely and appropriate behavioral health care can reduce effects of trauma and promote resilient PTG, some key lessons learned for managing mental health care from the disaster mental health management perspectives in Thailand will be addressed. Natural hazards have a tremendously negative impact on affected individuals. The negative impact can be exacerbated when a disaster strikes the community with underprepared infrastructure and limited structural support. Thailand was one of the countries with limited resources and infrastructural support for mental health care to the affected individuals. Well-prepared mental health management plans are essential to alleviate the impact of the disaster on individuals, families, communities, and the whole country.

Knowledge and skills in how trauma from a disaster can affect individuals and communities are essential to respond to the psychological needs that change over time and highly recommended by the World Health Organization (WHO; van Ommeren, Sexena, & Sarraceno, 2005). Traumatic stress due to having experienced a disaster may manifest in different levels of severity and symptomatology across times and developmental stages of the affected individuals (Kar & Bastia, 2006). Because not all affected individuals will develop psychological disorders after experiencing a disaster, early detection and interventions are imperative in determining the trajectory of the survivor's mental health recovery. Recognizing how and when these symptoms may occur in affected individuals may inform intervention strategies to increase the mental health effects of an intervention.

PTSD and depression are two of major psychiatric disorders that may occur following disaster survivors both in children and adults (Guo, He, & Qu, 2016; Kar & Bastia, 2006). This statistic is consistent with the prevalence found in a large study of 3,133 adult survivors from all 6 affected provinces in Thailand (Thavichachart et al., 2009), which reported that 33.60% were diagnosed with posttraumatic disorder (PTSD), 14.27% were diagnosed with depression, and 11.27% were diagnosed with both PTSD and depression. A longitudinal study also confirmed high rates of PTSD and depression at 2 months after the tsunami (33.6% and 25.7%) compared to at three years after the tsunami (2.6% and 18.7%) (Panyayong & Pengjantr, 2014).

Shortly after the tsunami hit Thailand (within 72 hours after), Thailand's Department of Mental Health (under guidance of Ministry of Public Health) took responsibility for coordinating services with other non-governmental organizations (NGOs) to provide mental health services to the affected communities. These services were organized in two phases – the initial phase and the recovery phase (Visanuyothin, Chakrabhand, & Bhugra, 2006). The initial phase focused on collecting information about immediate physical and mental health needs from the affected individuals by dispatching a mobile mental health team to the affected areas. The needs were collected from questionnaires as well as direct observations of behavioral and emotional symptoms (Visanuyothin et al., 2006). The main purpose of the assessments was not to pathologize the normal human response to such trauma but rather to keep records of the actual needs and strategize interventions to address these needs. Mental health symptoms during this time often could be classified as acute stress disorder, which may include symptoms similar to PTSD, especially dissociative amnesia, detachment, a subjective sense of numbing, a reduction in awareness of his or her surroundings that last within the first 30 days after the disaster (Bryant, 2006; Fullerton, Ursano, & Wang, 2004).

The recovery phase (after 72 hours) was focused on rescue and recovery tasks. The main tasks in this phase were to provide services for physical and mental health services in an integrated manner in order to reduce stigma and to increase therapeutic alliance and adherence (Visanuyothin et al., 2006). This phase drew an outpouring of national and international attention from NGOs and media outlets into the affected areas, which presented a particular challenge for the Department of Mental Health to streamline and organize the local mental health teams and the foreign humanitarian mental health workers to avoid conflicts, confusion, and culturally incongruent practices. In addition, during this period, the unprecedented systemic stressors, defined as "a loss or lack of essential community resources necessary for the effective functioning of citizens," became increasingly evident and continued to exacerbate mental health symptoms (Sattler, Asssanangkorncai, Moller, Kesavatana-Dohre, & Graham, 2014, p. 202). Systemic stressors found during this time included the lack of well-organized organizational structures and trained mental health professionals and paraprofessionals to be available to meet the physical and mental needs of the survivors who were displaced from home, lost family members and livelihood across the age groups (van Griensven et al., 2006). Risk and resilience factors associated with psychological problems and psychological recovery found during this phase included certain demographic factors (e.g., age, gender, and income), pre-tsunami mental health problems, and post-tsunami psychosocial support (Sattler et al., 2014).

There are lessons learned from studies of direct mental health services in Thailand after the tsunami. These lessons appear to be aligned with the principles recommended by WHO for disaster management (van Ommeren et al., 2005). First, the addition of a preparation phase is recommended in order to establish a cohesive organizational infrastructure to prepare for future disasters (Panyayong & Pengjuntr, 2006). During this phase, systemic capacity building should be emphasized in order to increase the level of preparedness. The capacity building is set to establish the Mental Health Operations Center under the leadership of the Department of Mental Health to act as a central commander to give an overall direction and oversee the work of mental health care teams from the supervision and evaluation center to the frontline operations centers in different provinces, which manage mobile mental health teams and volunteer teams based on affected provinces (Panyayong & Pengjuntr, 2006). For mobile mental health workers, trainings are being developed to ensure high-quality mental health interventions. In addition, the National Guidelines for Mental Health Intervention in Natural hazards were developed to provide essential, specific, and coherent information to all staff and volunteers involved to increase their readiness and preparedness for future disasters (Panyayong & Pengjuntr, 2006).

Second, extending the Recovery/Rehabilitation phase from two months after the disaster to three years after the disaster is recommended to ensure the continuity of mental health support for affected individuals (Panyayong & Pengjuntr, 2006; Visanuyothin et al., 2006). This recommendation is based on the growing understanding of the importance of early as well as continuing support. During this phase, changing the status from victims to survivors helps mobilize energy into rebuilding and revalidating meaning in life and reconnecting with their community (Visanuyothin et al., 2006). Indeed, one study of PTG of adult Thai survivors of the 2004 Southeast Asian earthquake tsunami six months after the disaster found the prevalence rates for interpersonal and intrapersonal PTG were 32% and 37%, respectively (Tang, 2007). Additional research of adult Thai survivors of the tsunami six months after the disaster found that the best predictors of positive adjustment were predisaster employment and frequent support seeking (Tang, 2006). Furthermore, a study of Norwegian adults present in Khao Lak, Thailand at the time of the 2004 tsunami assessed two and six years after the disaster found that posttraumatic stress and depression were negatively associated with quality of life, but that PTG may serve as a moderating factor in this relationship, such that PTG reduces stress and depression (Siqveland, Nygaard, Hussain, Tedeschi, & Heir, 2015).

Third, cultural competence strategies in postdisaster mental health care are essential (Thavichachart et al., 2009; Udomratn, 2008). The involvement of religious leaders and mind-body interventions as well as psychosocial support is highly recommended in the context of Thailand (Udomratn, 2008). After the tsunami, there were a number of foreign humanitarian mental health workers who brought Western interventions in order to offer help to the affected individuals. Most of these interventions tended to focus on the diagnosis of PTSD. However, there is a number of issues that must be considered when Westerners offer to provide these services in non-Western countries (Leitch, 2007). For example, the meaning of symptoms may differ across cultures, many cultures may have non-equivalent terms for PTSD symptoms, evidence is limited that Western treatments for PTSD are affective across cultures, and imported Western techniques may undermine traditional healing mechanisms (Silove & Bryant, 2006). Therefore, it is important for foreign mental health volunteers and workers to work closely with local mental health professionals, community members, and religious leaders in the community to design a culturally congruent mental health intervention.

It was clear that prior to the Indian Ocean Tsunami in 2004, the most devastating disaster in Thailand's contemporary history, Thailand's disaster preparedness for mental health care was not yet fully equipped with essential knowledge and skills to respond to a disaster of such magnitude that left such tremendous damages. To be more prepared for future disasters, Thailand's Department of Mental Health has several plans that are aligned with the WHO's 2005 recommendations via capacity building and organizational structures that give directions for different phases of responses, including the preparation phase, initial phase, and recovery/rehabilitation phase. Additional recommendations would likely include policy to advise that mental health programs for disaster survivors should focus upon active coping strategies, including support seeking, strengthening of support networks, and managing posttraumatic stress symptoms to increase positive adjustment (Tang, 2006). Psychologists should also note the potential impact of the type of exposure on posttraumatic stress and PTG. For instance, a study of survivors of a 2004 tsunami living in Stockholm six years after the tsunami found that those who experienced a combination of types of exposure (such as threat to life and bereavement) reported higher scores for both posttraumatic depreciation (negative changes in survivors' life view) and for PTG (Michélsen, Therup-Svedenlöf, Backheden, & Schulman, 2017).

Research based upon examination of survivors from the 2004 earthquake-tsunami also has implications and relevance for psychological interventions following other natural hazards elsewhere in SEA. For instance, a study of relationships among posttraumatic stress, resource loss, depression, social support, and PTG following a devastating earthquake in Central Java, Indonesia found that 95% of participants who were living in rural farming villages experienced either damage to homes or complete home destruction (Sattler, Claramita, & Muskavag, 2018). Severity of posttraumatic symptoms was associated with resource loss, depression, and fears about possible future earthquakes. PTG was associated with energy resource loss and social support, in keeping with conservation of resources stress theory (COR). COR states in part that persons are motivated to obtain valued resources (e.g., objects, conditions such as employment/marriage, personal characteristics, energy/time) that offer meaning, safety, and security) (Hobfoll, 2010). A study of Thai survivors of the 2004 tsunami also found that social support was associated with PTG (Sattler et al., 2018). The role of spirituality in PTG following natural hazards in SEA merits additional exploration as well. For instance, a study of survivors of the 2010 eruption of Mount Merapi in Java, Indonesia, found that both qualitative and quantitative data from this sample indicated that spirituality, such as through prayer, belief in God, compassion, and patience, transformed posttraumatic stress experiences into increases in PTG (Subandi, Achmad, Kurniati, & Febri, 2014). Adding further evidence for the contributing role of religiosity to PTG, a study of Filipino survivors of typhoon Haiyan found that generalized beliefs about the social world influenced the development of PTG (Nalipay, Bernardo, & Mordeno, 2016). The social axiom of religiosity refers to the concept that religion is a good force that contributes to positive psychological outcomes such as active coping, self-esteem, life satisfaction, and interpersonal harmony (Nalipay et al., 2016).

PTG AND RESILIENCE

Research over the past 20 years has established that many people can and do report personal growth, and some positive outcomes as well, as a result of their challenges and struggles after a trauma or other stressful life event (Bonanno, 2005; Calhoun & Tedeschi, 2006; McMillen, 1999). Indeed, the term PTG was coined to help articulate and conceptualize this idea. Calhoun and Tedeschi (2013), leading researchers of PTG, define it as "the experience of positive change that the individual experiences as a result of the struggle with a traumatic event" (p. 6). Calhoun and Tedeschi noted that interest in this concept increased in the 1990s, and that research on the topic increased with the development of measures and assessments regarding PTG, such as the Changes in Outlook Questionnaire, the Stress-Related Growth Scale, and the Posttraumatic Growth Inventory (Calhoun & Tedeschi, 2013). In addition to purely quantitative measures, psychologists and anthropologists have also examined the concept with mixed methods (e.g., Rich & Sirikantraporn, 2018) and with qualitative interviews and fieldwork (e.g., Ferrara, 2018; Sirikantraporn et al., 2018). Qualitative work is especially valuable in represented lived experiences and also in capturing free-response data that may be missed with forced choice quantitative survey items that are normed in the United States or Western Europe (Rich, 2016).

A number of studies from Europe and the United States suggest a certain set of common elements relating to PTG. These core parts of PTG include improved relating to others, seeing new possibilities in life, spiritual change, and increases in personal strength and appreciation of life (Calhoun & Tedeschi, 2013; Tedeschi & Calhoun, 1996). Though the earlier PTG studies were overwhelmingly conducted with Western participants, more recently scholars are beginning to examine PTG and related concepts such as resilience across cultures, using both quantitative and qualitative studies. The qualitative component is less common in much conventional psychology yet is particular suited for examining cultural nuances that may otherwise be missed (Rich, 2016). As examples of some of the relevant international PTG research, one may note work with samples from Australia (Shakespeare-Finch & Copping, 2006), Japan (Taku, Calhoun, Tedeschi, Gil-Rivas, Kilmer, & Cann, 2007), Cambodia (Sirikantraporn et al., 2018), India (Sirikantraporn, Rich, & Badaracco, 2018), the Philippines (Nalipay & Mordeno,

2018), and China (Ho et al., 2013), and from cancer survivors across cultures (Diaz, Cordova, & Spiegel, 2012).

Generally speaking, results from these international studies suggest that the five dimensions of PTG (personal strength, appreciation of life, relating to others, spiritual change, and new possibilities; Tedeschi & Calhoun, 1996) may or may not vary across cultures, especially for cultures in which the self is defined as individualistically oriented, compared to collectivistically oriented (Helgeson, Reynolds, & Tomich, 2006; Yang, 2003). For instance, some data suggest that in collectivistic cultures such as Haiti, India, and Cambodia, the importance of family and connections to religious communities and concepts may be especially salient in comparison to more individualistic cultures where focus on personal growth may be relatively more emphasized; thus connections to family and to faith traditions may be especially important for PTG in collectivistic cultures (Rich & Sirikantraporn, 2018).

Though research on PTG specifically in SEA is still a rarity, initial data (e.g., Sirikantraporn et al., 2018) indicate that in general the model is supported, as in the case of research with a contemporary sample in Cambodia, though there are some cultural nuances. For instance, although Calhoun and Tedeschi (2013) described five core factors of PTG (personal strength, relating to others, new life possibilities, appreciation of life and spirituality), four categories of PTG (personal growth, interpersonal growth, life transitions, and changed role of religion/spirituality) emerged from qualitative interviews conducted with a Cambodian sample (Sirikantraporn et al., 2018). In the Cambodia case, in the 40 years since the peak of the Khmer Rouge era, many cultural studies, including scholarship in cultural psychiatry and in medical anthropology, have explored the nuances of trauma and trauma responses as experienced among Cambodians as a result of the Khmer Rouge (Becker, Beyene, & Ken, 2000; Hinton, Hinton, Pich, Loeum, & Pollack, 2009; Hinton, Howes, & Kirmayer, 2008; Ratliff, 1997; Van Schaack et al., 2011). Though such studies are valuable, in order to gain a full picture of trauma study in context, it is vital to examine the local nuances and meanings of growth and positive changes after trauma. For instance, previous research reveals a number of localized, and even unique cultural responses in the Cambodian context which do not occur in majority culture "Western" samples such as from the United States. As examples, salient culturally specific trauma-related symptoms of survivors of the Khmer Rouge, include *kyol goeu* (wind overload), which causes one to become unconscious, fall to the ground, turn pale, and sweat (Hauff, 2001), sleep paralysis ("the ghost pushes you down"), nightmares, and panic attacks (Hinton, Pich, Chhean, & Pollack, 2005). Also notable is the local Cambodian trauma phenomenon of broken courage (*baksbat*), which is somewhat similar to PTSD (Chhim, 2013). *Baksbat* is the idiom of distress experienced by Khmer persons as a lack of trust in others, submissiveness, feeling fearful, and being mute and deaf (Chhim, 2013). Many of these culturally specific symptoms have a strong interpersonal component. This interpersonal nature, as for sleep paralysis and *baksbat,* may partially explain why healing practices found especially helpful for Cambodian trauma survivors are often interpersonal by nature. Some such practices among Cambodian trauma survivors are making merits to the deceased, holding a festival for loved ones, and healing the wider community (Agger, 2015).

In sum, in the contemporary Cambodia context (Sirikantraporn et al., 2018), the Tedeschi and Calhoun (1996) model of PTG is generally supported, but with some cultural variation. For instance, data indicate participants developed a maturing perspective, with a sense of increased personal strength, and a shift from a powerless to an empowered identity. Furthermore, qualitative interview data indicate a greater number and a greater quality of connections with others, and an activated desire to help others (Sirikantraporn et al., 2018). Finally, the ability to find meaning in suffering in a future-oriented manner promoted posttraumatic growth after traumatic events. One aspect of PTG that was not in the original Tedeschi and Calhoun (1996) model but which emerged in the Cambodian sample is the concept of karma. Karma is a core component in Khmer Buddhism, though complex to fully grasp, in general it relates to the cause–effect of one's actions, and to the idea that one deserves things that happen to them because of one's past actions, including in past lives (Smith-Hefner, 1999). After the Khmer Rouge time of Pol Pot, ample research has aimed to explore the role of Karma in individuals' well-being and understanding for life events (e.g., Coggan, 2015; Harris, 2008; Ratliff, 1997). Given the fate of many of the religious leaders as well as intellectuals during the Pol Pot regime, it is not surprising that today's Cambodians' understanding of karma is in transition (Ledgerwood, Ebihara, & Mortland, 1994). The unique history of Cambodia with respect to its neighbors such as Thailand, also illustrate how there are both commonalities (e.g., Theraveda Buddhism) and differences (e.g., history of genocide) in the region, making it challenging to make effective broad statements, and pointing to a critical need for further quality studies of PTG and resilience in the region, to best reflect current, local variations.

CONCLUSION

This chapter has aimed to illustrate how concepts from positive psychology and strengths-based research, can sensibly be applied and internationalized to issues focused on disaster, crisis, and trauma in the SEA context, by emphasizing the applications for resilience and PTG. The scope, scale, and diversity of SEA precludes review of every type of traumatic event or disaster, as manifested in every nation and village in the region, yet lessons from research may still be gleaned from close examination of the examples and research literature presented here. Evidence suggests much that is worthwhile in the original Western research on PTG and resilience (Calhoun & Tedeschi, 2014) and that is relevant for SEA (e.g., Rich & Sirikantraporn, 2018), yet further research is needed with SEA populations to ensure cultural nuances are adequately assessed and addressed. Issues relating to positive coping styles and interactions with language, culture, religion, economics, political realities, and systems as well as social standing require further investigation to ensure that the most effective and appropriate interventions may be developed and implemented for the local context.

REFERENCES

Agger, I. (2015). Calming the mind: Healing after mass atrocity in Cambodia. *Transcultural Psychiatry*, 1–18. doi:10.1177/1363461514568336t

Becker, G., Beyene, Y., & Ken, P. (2000). Memory, trauma, and embodied distress: The management of disruption in the stories of Cambodians in exile. *Ethos*, *28*(3), 320–345.

Bonanno, G. A. (2005). Clarifying and extending the construct of adult resilience. *American Psychologist*, *60*(3), 265–267

Brown, D. E. (1991). *Human universals*. Boston, MA: McGraw Hill.

Bryant, A. R. (2006). Acute stress disorder. *Psychiatry*, *5*(7), 238–239.

Calhoun, L. G., & Tedeschi, L. G. (2014). *Facilitating posttraumatuc growth*. Mahwah, NJ: Routledge.

Calhoun, L. G., & Tedeschi, R. G. (Eds.). (2006). *Handbook of posttraumatic growth: Research and practice*. New York, NY: Psychology Press.

Calhoun, L. G., & Tedeschi, R. G. (2013). *Posttraumatic growth in clinical practice*. New York, NY: Routledge.

Chandler, D. (2007). *A history of Cambodia* (4th ed.). Philadelphia, PA: Westview.

Chhim, S. (2013). Baksbat (broken courage): A trauma-based cultural syndrome in Cambodia. *Medical Anthropology*, *32*(2), 160–173.

Coggan, P. (2015). *Spirit worlds: Cambodia, the Buddha, and the Naga*. Oxford: John Beaufoy Publishing Limited.

Diaz, M., Cordova, M. J., & Spiegel, D. (2012). Posttraumatic growth in cancer patients across cultures. In L. Grassi & M. Riba (Eds.), Clinical psycho-oncology: An international *perspective* (pp. 211–222). New York, NY: John Wiley & Sons.

Farmer, P. (2011). *Haiti after the earthquake*. New York, NY: Public Affairs.

Ferrara, N. (2018). *In pursuit of impact: Trauma and resilience-informed policy development*. Lanham, MD: Lexington Books.

Fullerton, C. S., Ursano, R. J., & Wang, L. (2004). Acute stress disorder, posttraumatic stress disorder, and depression in disaster or rescue workers. *The American Journal of Psychiatry*, *161*, 1370–1376. doi:10.1176/appi.ajp.161.8.1370

Guo, J., He, H., & Qu, Z. (2016). Post-traumatic stress disorder and depression among adult survivors 8 years after the 2008 Wenchuan earthquake in China. *Journal of Affective Disorders*, *210*, 27–34.

Harris, I. (2008). *Cambodian Buddhism: History and practice*. Honolulu, HI: University of Hawai'i Press.

Hauff, E. (2001). Kyol Goeu in Cambodia. *Transcultural Psychiatry*, *38*, 468–473.

Helgeson, V. S, Reynolds K. A., & Tomich, P. L. (2006). A meta-analytic review of benefit finding and growth. *Journal of Consulting and Clinical Psychology*, *74*, 797–816.

Henrich, J., Heine, S. J., & Norenzayan, A. (2010). The weirdest people in the world? *Behavioral and Brain Sciences*, *33*(2–3), 61–83. http://dx.doi.org/10.1017/S0140525X0999152X

Hinton, A. L. (2002). *Annihilating difference: The anthropology of genocide*. Berkeley, CA: University of California Press.

Hinton, A. L. (2004). *Why did they kill?: Cambodia in the shadow of genocide*. Berkeley, CA: University of California Press.

Hinton, D. E., Hinton, A. L., Pich, V., Loeum, J. R., & Pollack, M. H. (2009). Nightmares among Cambodian refugees: The breaching of concentric ontological security. *Cultural Medical Psychiatry*, *33*(2), 219–265.

Hinton, D. E., Howes, D., & Kirmayer, L. J. (2008). Toward a medical anthropology of sensations: Definitions and research agenda. *Transcultural Psychiatry*, *45*(2), 142–162.

Hinton, D. E., Pich, V., Chhean, D., & Pollack, M. H. (2005). "The ghost pushes you down": Sleep paralysis-type panic attacks in a Khmer refugee population. *Transcultural Psychiatry*, *42*(1), 46–77.

Ho, S. M., Law, L. S., Wang, G. L., Shih, S. M., Hsu, S. H., & Hou, Y.C. (2013). Psychometric analysis of the Chinese version of the Posttraumatic Growth Inventory with cancer patients in Hong Kong and Taiwan. *Psycho-Oncology*, *22*, 157–179.

Hobfoll, S. E. (2010). Conservation of resources theory: Its implication for stress, health, and resilience. In S. Golkman (Ed.), *The Oxford handbook of stress, health, and coping*. New York, NY: Oxford. doi:10.1093/oxfordhb/9780195375343.013.0007

Hyndman, J. (2011). *Dual disasters: Humanitarian aid after the 2004 Tsunami*. Kumarian Press.

Jarl, J., Cantor-Graae, E., Chak, T., Sunbaunat, K., & Larsson, C. A. (2015). Trauma and poor mental health in relation to economic status: The case of Cambodia 35 years later. *PLoS one*, *10*(8), e0136410. doi:10.1371/journal.pone.0136410

Joseph, S., & Linley, P. A. (Eds.). (2008). *Trauma, recovery, and growth: Positive psychological perspectives on posttraumatic stress*. Hoboken, NJ: John Wiley.

Kar, N., & Bastia, B. K. (2006). Post-traumatic stress disorder, depression and generalized anxiety disorder in adolescents after a natural disaster: a student of comorbidity. *Clinical Practice Epidemiology in Mental Health*, *2*(17), 1–7.

Kierman, B. (2008). The Pol Pot regime: Race, power, and genocide in Cambodia under the Khmer Rouge, 1975-1979 (3rd ed.). New Haven, CT: Yale University Press.

Kristof, N. D., & WuDunn, S. (2009). *Half the sky: Turning oppression into opportunity for women worldwide*. New York, NY: Vintage Random House.

Kuriansky, J. (Ed.). (2007). *Beyond bullets and bombs: Grassroots peacebuilding between Israelis and Palestinians*. Westport, CT: Praeger.

Kuriansky, J. (Ed.). (2016). *The psychosocial aspects of a mass epidemic*. Santa Barbara, CA: ABC-CLIO.

Ledgerwood, J., Ebihara, M.M., & Mortland, C. (1994). Introduction. In M. M. Ebihara, C.A. Morland, & J. Ledgerwood (Eds.), *Cambodian culture since 1975: Homeland and exile* (pp. 1–26). Ithaca, NY: Cornell University Press.

Leitch, M. L. (2007). Somatic experiencing treatment with tsunami survivors in Thailand: Broadening the scope of early intervention. *Traumatology*, *13*(3), 11–20.

Leong, F. T. L., Pickren, W. E., Leach, M. M., & Marsella, A. J. (Eds.). (2012). *Internationalizing the psychology curriculum in the United States*. New York, NY: Springer.

Lomas, T. (2018). *Translating happiness: A cross-cultural lexicon of well-being*. Cambridge, MA: MIT Press.

Mathews, G., & Izquierdo, C. (Eds.). (2009). *Pursuits of happiness: Well-being in anthropological perspective*. New York, NY: Berghahn Books.

McMillen, J. C. (1999). Better for it: How people benefit from adversity. *Social Work*, *44*(5), 455–468.

Michélsen, H., Therup-Svedenlöf, C., Backheden, M., & Schulman, A. (2017). Posttraumatic growth and depreciation six years after the 2004 tsunami. *European Journal of Psychotraumatology*, *8*(1), 1–10.

Nalipay, M. J. N., Bernardo, A. B. I., & Mordeno, I. G. (2017). Posttraumatic growth in survivors of a natural disaster: The role of social axioms of religiosity, reward for application, and social cynicism. *Journal of Positive Psychology*, *12* (4), 342–353.

Nalipay, M. J. N., & Mordeno, I. G. (2018). Positive metacognitions and meta-emotions as predictors of posttraumatic stress disorder and posttraumatic growth in survivors of a natural disaster. *Journal of Loss & Trauma*, *23*(5), 381–394.

Oishi, S. (2010). Culture and well-being: Conceptual and methodological issues. In E. Diener, J. F. Helliwell, & D. Kahneman (Eds.), *International differences in well-being* (pp. 34–69). New York, NY: Oxford University Press.

Panyayong, B., & Pengjuntr, W. (2006). Mental health and psychosocial aspects of disaster preparedness in Thailand. *International Review of Psychiatry*, *18*(6), 607–614.

Panyayong, B., & Pengjuntr, W. (2014). PTSD, depression and anxiety among tsunami affected people in Thailand: A population-based longitudinal study. *Natural Sciences*, *6*(2014), 114–120.

Paton, D., Violanti, J. M., & Smith, L. M. (Eds.). (2002). *Promoting capabilities to manage posttraumatic stress: Perspectives on resilience*. Springfield, IL: Charles C. Thomas.

Peterson, C. (2006). *A primer in positive psychology*. New York, NY: Oxford University Press.

Ratliff, S. R. (1997). *Caring for Cambodian-Americans: A multidisciplinary resource for the helping professions*. New York, NY: Garland Publishing.

Rhoades, G. F., & Sar, V. (Eds.). (2005). *Trauma and dissociation in a cross-cultural perspective*. New York, NY: Routledge.

Rich, G. (2015, August). *Internationalizing psychology in Cambodia: Lessons learned and future directions*. Paper presented at the annual meeting of the American Psychological Association. Toronto, Canada.

Rich, G. (2016). The promise of qualitative inquiry for positive psychology: Diversifying methods. *Journal of Positive Psychology*, *12*(3), 220–231. doi:10.1080/17439760.2016.1225119

Rich, G. (2017). Positive psychology and humanistic psychology: Evil twins, sibling rivals, distant cousins or something else? *Journal of Humanistic Psychology*. doi:10.1177/0022167817698820

Rich, G., Gielen, U., & Takooshian, H. (Eds.). (2017). *Internationalizing the teaching of psychology*. Charlotte, NC: IAP- Information Age Publishing.

Rich, G., Jaafar, J., & Barron, D. (Eds.). (In press). *Psychology in Southeast Asia*. New York, NY: Routledge.

Rich, G., & Sirikantraporn, S. (Eds.). (2018). *Human strengths and resilience: Developmental, cross-cultural, and international perspectives (foreword by Chris Stout)*. Lanham, MD: Lexington Books.

Royal University of Phnom Penh (RUPP). (2012). *Cambodia Mental Health Survey*. Phnom Penh: RUPP Department of Psychology.

Saleeby, D. (Ed.). (2008). *The strengths perspective in social work practice* (5th ed.). Boston, MA: Allyn & Bacon.

Sattler, D. N., Assanangkornchai, S., Moller, A. M., Kesavatana-Dohrs, W., & Graham, J. M. (2014). Indian ocean tsunami: Relationships among posttraumatic stress, posttraumatic growth, resource loss, and coping at 3 and 15 months. *Journal of Trauma & Dissociation, 15*, 219–239.

Sattler, D. N., Claramita, M., & Muskavage, B. (2018). Natural disasters in Indonesia: Relationships among posttraumatic stress, resource loss, depression, social support, and posttraumatic growth. *Journal of Loss & Trauma, 23*(5), 351–365.

Seligman, M. E. P. (2002). *Authentic happiness*. New York, NY: Free Press.

Seligman, M. (2011). *Flourish*. New York, NY: Free Press.

Seligman, M. E. P. & Csikszentmihalyi, M. (2000). Positive psychology: An introduction. *American Psychologist, 55*, 5–14.

Shakespeare-Finch, J. E., & Copping, A. (2006). A grounded theory approach to understanding cultural differences in posttraumatic growth. *Journal of Loss and Trauma, 11*, 355–371.

Silove, D., & Bryant, R. (2006). Rapid assessments of mental health needs after disasters. *Journal of the American Medical Association, 296*(5), 576–578.

Siqveland, J., Nygaard, E., Hussain, A., Tedeschi, R. G., & Heir, T. (2015). Posttraumatic growth, depression and posttraumatic stress in relation to quality of life in tsunami survivors: A longitudinal study. *Health & Quality of Life Outcomes, 13*(1), 1–8.

Sirikantraporn, S., Rich, G., & Badaracco, J. (2018). The concept of posttraumatic growth in a sample of undergraduates from India: A mixed methods study. In G. Rich & S. Sirikantraporn (Eds.), *Human strengths and resilience: Developmental, cross-cultural, and international perspectives* (pp. 1–20). Lanham, MD: Lexington Books.

Sirikantraporn, S., Rich, G., & Jafari, N. (2018). The concept of posttraumatic growth in a Cambodian sample: A grounded theory study. In G. Rich & S. Sirikantraporn (Eds.), *Human strengths and resilience: Developmental, cross-cultural, and international perspectives* (pp. 39–58). Lanham, MD: Lexington Books.

Smith-Hefner, N. J. (1999). *Khmer Americans: Identity and moral education in a diasporic community*. Berkeley, CA: University of California Press.

Subandi, M.A., Achmad, T., Kurniati, H., & Febri, R. (2014). Spirituality, gratitude, hope and post-traumatic growth among the survivors of the 2010 eruption of Mount Merapi in Java, Indonesia. *Australasian Journal of Disaster and Trauma Studies, 18*(1), 19–26.

Takooshian, H., Gielen, U., Plous, S., Rich, G., & Velayo, R. (2016). Internationalizing undergraduate psychology education: Trends, techniques, and technologies. *American Psychologist, 71*(2), 136–147.

Tang, C. S. K. (2006). Positive and negative postdisaster psychological adjustment among adult survivors of the Southeast Asian earthquake-tsunami. *Journal of Psychosomatic Research, 61*(5), 699–705.

Tang, C. S. K. (2007). Posttraumatic growth of Southeast Asian survivors with physical injuries: Six months after the 2004 Southeast Asian earthquake-tsunami *Australasian Journal of Disaster and Trauma Studies, 1*, 1–12.

Taku, K., Calhoun, L. G., Tedeschi, R. G., Gil-Rivas, V., Kilmer, R. P., & Cann, A. (2007). Examining posttraumatic growth among Japanese university students. *Anxiety Stress Coping, 20*(4), 353–367.

Tedeschi, R. G., & Calhoun, L.G. (1996). The posttraumatic growth inventory: Measuring the positive legacy of trauma. *Journal of Traumatic Stress, 9*(3), 455–471.

Thavichachart, N., Sookjaroen, T., Worakul, P., Kanchanatawan, B., Suppapitiporn, S., Sukoltapirom na Pattalung, A., & Roomruangwong, C. (2009). Posttraumatic mental health establishment of the tsunami survivors in Thailand. *Clinical Practice and Epidemiology in Mental health*, *5*(11), 1–9.

Theron, L., Liebenberg, L., & Ungar, M. (Eds.). (2015). *Youth resilience and culture: Commonalities and complexities (Cross-cultural advancements in positive psychology)*. New York, NY: Springer.

Udomratn, P. (2008). Mental health and the psychosocial consequences of natural disasters in Asia. *International Review of Psychiatry*, *20*(5), 441–444.

Van der Kolk, B. (2015). The body keeps the score: Brain, mind, and body in the healing of trauma. New York, NY: Viking.

van Griensven, F., Chakkraband, M. L. S., Thienkrua, W., Pengjuntr, W., Cardozo, B. L., & Tantipiwatanaskul, P. (2006) Rapid assessment of post-tsunami mental health problems among adults in southern Thailand. *JAMA*, *296*, 537–548.

van Ommeren, M., Sexena, S., & Sarraceno, B. (2005). Mental and social health during and after acute emergencies: Emerging consensus? Round Table. *Bulletin of the World Health Organization*, *83*, 71–75.

Van Schaack, B., Reicherter, D., & Chhang, Y. (Eds.). (2011). *Cambodia's hidden scars: Trauma psychology in the wake of the Khmer Rouge*. Phnom Penh: Documentation Center of Cambodia.

Visanuyothin, T., Chakrabhand, S., & Bhugra, D. (2006). Tsunamic and mental health in Thailand. *International Review of Psychiatry*, *18*(3), 271–273.

Weiss, T., & Berger, R. (Eds.). (2010). *Posttraumatic growth and culturally competent practice: Lessons learned from around the globe*. Hoboken, NJ: John Wiley and Sons.

Wong, P. (Ed.). (2012). *The human quest for meaning* (2nd ed.). New York, NY: Taylor and Francis.

Yang, K. S. (2003). Beyond Maslow's culture-bound linear theory: A preliminary statement of the double-Y model of basic human needs. *Nebraska Symposium on Motivation*, *49*, 175–255.

Zoma, M., & Gielen, U. P. (2015). How many psychologists are there in the world? *International Psychology Bulletin*, *19*(1), 47–50.

PART IV

CONCLUSION

CHAPTER 10

CHALLENGES AND PROSPECTS IN PROMOTING RESISTANCE, RESILIENCE, AND RECOVERY IN SOUTHEAST ASIA

Ma. Regina M. Hechanova and Lynn C. Waelde

ABSTRACT

Advances in disaster prevention and mitigation in Southeast Asia (SEA) have increasingly included plans for mental health and psychosocial support. However, substantial challenges remain, particularly in the areas of (a) disaster communication and preparedness, (b) institutionalized disaster education, (c) culturally adapted and evidence-based tools and interventions, (d) developing capacities and caring for disaster responders, and (e) enabling collective resilience. In addition, the impacts of poverty, lack of access to education, and other forms of marginalization result in less resources to prepare for hazardous event and increased vulnerability to environmental hazards for SEA countries. These issues highlight the need for SEA governments to address deeply rooted human development issues that put communities at risk and heighten vulnerabilities of SEA populations.

Keywords: Mental health and psychosocial support; disaster preparation; Southeast Asia; disaster education and training; culturally adapted interventions; disaster responders

---

Resistance, Resilience, and Recovery from Disasters: Perspectives from Southeast Asia
Community, Environment and Disaster Risk Management, Volume 21, 161–171

ISSN: 2040-7262/doi:10.1108/S2040-726220200000021010

Given its vulnerability to disasters, disaster prevention and mitigation is not only important, it is vital in Southeast Asia (SEA). The region's location puts many of the countries in SEA within the Pacific Ring of Fire. As such, countries in SEA are prone to tsunamis, earthquakes, and volcanic eruptions (Samphantharak, 2014). Beyond natural hazards, the region has also had its share of human-made disasters include war, terrorism, political and/or social instability, high rates of poverty, and often very limited access to resources (Rich & Sirikantraporn, 2020).

It is important to note the recent improvements in disaster prevention and mitigation across the region, such as geographical mapping of existing resources, risk areas, and vulnerable groups, and performance of local risk assessments, and the establishment of early warning systems and response/evacuation plans (Labaria, Acosta, & Gotangco, 2020). Another positive development in the region is the emergence of plans, structures, and policies that embed mental health and psychosocial support (MHPSS) postdisaster. Thailand was one of the first country in the region to do this through its National Health Act of 2007 and a Mental Health Act was passed in 2008. Thailand's National Disaster Risk Management Plan (2015) delegates mental health programs to the Ministry of Public Health, and psychological rehabilitation through social interactions to the Ministry of Social Development and Human Security (Labaria et al., 2020). Indonesia followed suit with its Health Law number 36/2009 (World Health Organization (WHO), 2011), and a Mental Health Law was passed in 2014 (Yusuf, 2016). In the Myanmar Action Plan on Disaster Risk Reduction (2017), the Ministry for Health and Sports is mandated to oversee health sector preparedness for response and recovery, which includes psychosocial support. In the Philippines, the National Disaster Risk Reduction and Management Plan for 2011–2028 delegated psychosocial disaster response efforts to the Department of Health (Office of Civil Defense, 2011).

Despite these improvements, however, there are still many gaps common to SEA countries. In this chapter, we summarize factors that contribute to these gaps and highlight possible areas for future work. We focus the challenges and prospects in terms of five themes: (a) disaster communication and preparedness, (b) institutionalized disaster education, (c) culturally adapted and evidence-based tools and interventions, (d) developing capacities and caring for disaster responders, and (e) enabling collective resilience.

DISASTER COMMUNICATION AND PREPAREDNESS

The past decade has seen improvements in information, education, and communication for raising awareness and encouraging public participation in disaster response in SEA. Drills are an important component of preparedness because they raise awareness for and build resilience (Allen, 2006; Karanci, Aksit, & Dirik, 2005). Muttarak and Pothishiri (2013), who conducted their study in risk areas in coastal Thailand, reported that households who participated in disaster trainings were more likely to undertake precautionary measures. However,

Labaria et al. (2020) also suggested that drills can be enhanced to go beyond what to do or where to go and incorporate possible scenarios. These kinds of simulations can build citizens' positive cognition and self-efficacy in problem-solving and decision-making.

However, not all disaster information campaigns are equal. A study in Aceh, Indonesia reported that the disaster education was ineffective and suggested that this could be due to lack of understanding of the local culture of Aceh (i.e., belief that tsunami is a punishment from God, and the use of religion to deal with pain and loss) (Ismail, Suwannapong, Howteerakul, Tipayamongkholgul, & Apinuntavech, 2016). These findings indicate that effectiveness is best achieved through contextualization within the specific culture of a community (Labaria et al., 2020). As Labaria et al. suggested, this is where communities can be harnessed through participatory mapping and determining appropriate early warning systems. If warning devices like sirens and blare horns cause panic to some groups, then alternative approaches such as telephone trees or even door-to-door warnings can be used. In addition, De Perez et al. (2015) asserted that it is vital to consult local experts who understand the context and can suggest effective disaster information campaigns. They also suggested that importance of planning interventions along a seasonal calendar.

INSTITUTIONALIZED DISASTER EDUCATION

In this volume, Roman Hoffmann and Daniela Blecha highlighted the importance of embedding disaster education in schools. They cited the promising example of Association of Southeast Asian Nations' (ASEAN's) "*Safe Schools Initiative*" (ASSI) that promotes the inclusion of disaster risk reduction in national curricula across eight South East Asian countries, including Cambodia, Indonesia, Lao PDR, Malaysia, Myanmar, Philippines, Thailand, and Vietnam (ASEAN 2018). It consists of hands-on safety preparedness lessons for students and teachers to equip them with a basic understanding of disaster risk reduction plans and how to prepare for and act during emergencies.

Hoffmann and Blecha (2020) contended that the effectiveness of disaster education depends on education and socioeconomic factors such as access to communication, lack of integration with the communities, and exclusion. Hoffmann and Muttarak (2017) also suggested the importance of non-formal or social learning such as enabling community members to reflect on their experiences and undertake actions to prepare for future disasters. Another notable example for non-formal education is Thailand's Tsunami Museum that has become a means for children to learn about disasters and appropriate responses to them (Labaria et al., 2020).

Given the scarcity of resources, harnessing technology and informal education is one way to reach more people in low-resource communities in SEA. Hoffmann and Blecha (2020) suggested that creative educational materials such as games, documentary and short videos, songs, puppetry and improvization theater are practical approaches can help fostering the disaster preparedness and analytic and problem-solving skills among the learners.

The "Child Centered Climate Change Adaptation (4CA)" program developed child-friendly education materials, such as animated films, cartoons, books, posters, and games to build knowledge of climate change related risks. Photography was used by asking children to identify, map, and build their knowledge of climate change and disaster risk reduction (Plan International & Australian Aid, 2019) .

Following the 2011 floods in Thailand, UNESCO in 2014 launched a mobile game application "Sai Fah: The Flood Fighter" that features a fictional girl Tanah who learns how to prepare to and protect herself from tsunamis and earthquakes. By navigating interactive challenges, users learn key concepts of disaster risk reduction and enhance their hazard awareness. Hoffmann and Blecha (2020) suggested that given the rise in technology use, particularly among children and adolescents, learning through games may be more effective than formal classroom teaching.

CULTURALLY ADAPTED AND EVIDENCE-BASED TOOLS AND INTERVENTIONS

The Interagency Standing Committee (IASC, 2007) guidelines on MHPSS in emergency settings highlight the need for culturally appropriate interventions. Manaois, Tabo-Corpuz, and Heise (2020) cited evidence that culturally adapted programs and interventions result in better engagement and retention rates and produce positive outcomes. A positive development in SEA is the cultural adaptation of Psychological First-Aid (PFA) that has been endorsed by the WHO as the MHPSS response during emergencies. Manaois et al. (2020) review of PFA in SEA highlighted a number of unique approaches including the incorporation of indigenous frameworks, mindfulness, and using groups and open-space methods in problem-solving. However, an article by Landoy, Hechanova, Ramos, and Kintanar (2015) suggested a number of challenges in delivering PFA in low-resource communities in the Philippines. These include the limited number of individuals trained on PFA, the lack of protocol for the safe monitoring and follow-up of people with severe psychopathology, the lack of conducive venues to conduct PFA, and difficulties in linking the survivors with the appropriate services because of inadequate disaster response system. Manaois et al. (2020) also highlighted the needs for more robust evaluations of PFA and its adaptations, increasing knowledge sharing on the use and adaptation of PFA, scaling up capacity building, and institutionalizing PFA in disaster response systems.

However, PFA is only meant as a first-response intervention and there is a need for mental health interventions for those who may need more assistance during the recovery phase. The World Health Organization (2016) developed an intervention for low-resource contexts called Problem Management Plus. It consists of five modules that build stress management skills, strengthen problem-solving, encourage positive behaviors, and strengthen social support. Although it was has been found to reduce anxiety and depressive symptoms in Pakistan (Rahman et al., 2016), it has not been adapted for SEA.

A resilience intervention that emerged after Typhoon Haiyan in the Philippines is called *Katatagan* (translated to mean forbearance or strength). The intervention contains six modules designed to build adaptive coping and self-efficacy among survivors during the recovery phase of a disaster. Its first module focuses on survivors' identifying their strengths to build their self-efficacy. The next modules focus on building survivors' capacity to manage negative their reactions as well as managing negative thoughts and emotions. The last modules focus on encouraging positive activities, seeking solutions and social support, and planning for the future. Using cognitive-behavioral theory, the modules use creative and interactive methods and indigenous symbols and metaphors. Moreover, the modules were designed so they could be delivered by lay facilitators. Studies evaluating *Katatagan* reported that it decreases posttraumatic stress symptoms, anxiety, and depressive symptoms, and increases coping self-efficacy of survivors (Hechanova et al., 2018; Hechanova, Waelde, Docena, et al., 2015; Hechanova, Waelde, & Ramos, 2016).

Albeit promising, there is a need for more culturally appropriate focused, non-specialized interventions in SEA. In addition, most evaluation studies of disaster interventions in SEA are case studies, qualitative studies, or one sample pre and post-test studies without randomized controlled trial designs (Seyle, Widyatmoko, & Silver, 2013). In addition, a critical gap appears to be culturally adapted tool for assessing psychosocial distress (Labaria et al., 2020). The lack of local tools and rigor in evaluation may be due to the lack of appreciation of research, the lack of research resources, and the lack of competencies in design and evaluation of interventions. Hence, there is also a need to strengthen the capability of local mental health professionals to design and evaluate interventions as well as resources to enable to do so.

DEVELOPING CAPACITIES OF AND CARING FOR DISASTER RESPONDERS

Beyond formal disaster education, a critical gap in SEA is health professionals who can provide postdisaster MHPSS. Unlike developed economies where universities offer degree programs on disasters risk reduction and management, formal education on this disaster response is scarce in SEA. A study of community nurses in seven Asia-Pacific countries revealed low-to-moderate levels of disaster knowledge, skills, and preparedness (Usher et al., 2015). This finding led to calls to embed disaster preparedness as part of the nursing curriculum (Juanita, Suratmi, & Maghfiroh, 2018; Usher et al., 2015). On a positive note, at Ateneo de Manila University, a course on disaster and its effect on mental health is also being offered as an elective for undergraduate students in psychology and those who minor in humanitarian action (Labaria et al., 2020). However, there is a need to scale-up these initiatives to embed formal education in the curricula of health professionals.

Even as it is important to build the capacity of health professionals, the reality is that there is a shortage of them in SEA (van Griensven et al., 2006; Hechanova et al., 2015). Studies suggest that Asians are more likely to seek help from family, friends, community leaders, and local healers than from professionals (Hechanova & Waelde, 2017; Tuliao, 2014). Given the reluctance to seek help, family and peer support and the presence of volunteers from church and non-profit organizations are protective factors. Given the dearth of mental health professionals and the salience of spirituality in SEA, it is important for disaster responders to collaborate with local healers and faith-based professionals. Aten, O'Grady, Milstein, Boan, and Schruba (2014) explained that spiritual leaders are often "gatekeepers" in their communities. As such, they can identify community needs, be tapped to provide MHPSS and refer those who need more specialized care to professionals.

Beyond face-to face capacity building, technology can be harnessed to provide non-formal education for humanitarian workers. Online technologies such as "Disaster Ready Initiative," provides online professional development resources available to relief and development workers and volunteers (Cornerstone on Demand Foundation 2019). In the Philippines, a partnership between the Knowledge Channel Foundation, Ateneo de Manila University, the Psychological Association of the Philippines, and the Department of Education produced a four-part video series to help educators in disaster stricken areas recover from trauma and prepare them to provide PFA to students and community members (https://www.knowledgechannel.org/archives/4132).

Beyond capacitating disaster responders, an often forgotten aspect in disaster management is caring for responders. This is even more critical in SEA for a number of reasons. The first is that given the dearth of health professionals, majority of responders are non-traditional responders, such as faith workers, non-government organizations, and teachers. Secondly, given the frequency of disasters in SEA, responders are likely to come from disaster-hit areas and are primary survivors as well. Even when they are not primary survivors, responders may also have family members or very close friends who were affected by the disaster, making them secondary survivors. Exposure to psychologically disturbing elements such as dead bodies, exposure to toxic chemicals and communicable diseases, enduring inadequate sleeping arrangements, insufficient equipment, and fatigue also put them at risk (Alexander & Klein, 2009). In addition, responders are often caught in the middle of political crossfires conflict or entangled in the problems of their respective organizations that adds to the cumulative stress and may lead to compassion fatigue or even vicarious trauma (Guilaran & Nguyen, 2020).

The challenge in caring for responders, however, is that they hesitate to ask for help. Gender norms that men should be strong and need to endure pain predisposes responders to vicarious trauma (Agbayani et al., 2019). Given these obstacles, Guilaran and Nguyen (2020) suggested that an important intervention for responders is to teach them self-care through regular exercise, proper diet, routine medical appointments, and adequate sleep. This also means encouraging them to connect with people within and outside the organization, having a hobby or an enjoyable activity outside disaster work and finding meaning in their work. Recent work in the Philippines showed that use of *Inner Resources for Stress*, a mindfulness and mantra training program delivered through a brief workshop and an eight-week home study program, was associated with pre/post decreases in depression and was perceived as credible and useful by disaster workers (Waelde, Hechanova, Ramos, Macia, & Moschetto, 2018). Given the role of spirituality in SEA, prayer or simply having quiet time, may be other ways to encourage self-care.

ENABLING COLLECTIVE RESILIENCE

Whether for survivors or responders, mental health outcomes are influenced by other elements of a larger system and thus MHPSS cannot be viewed in a vacuum. At the base of the IASC hierarchy are services to ensure that basic needs are met – adequate food and shelter and healthcare and safety. The second tier in the pyramid focuses on community and family support whether in the form of emotional or instrumental support, or efforts for unifying families. Thus, organizations and communities have a critical role to play.

Particularly in developing economies where inadequate resources, a lack of communication infrastructure, turf wars between government institutions, and corruption can derail disaster response from government, it is communities that have immediate impact on survivors. Citing examples of disasters all over the world, Quarantelli's (1999) review highlighted how in the absence of pre-existing structures, what emerges is the solidarity and altruism of groups of people in sharing resources, helping injured people, organizing collectively, and putting their own lives at risk.

Especially in collective cultures, communities play crucial roles during disasters. Unfortunately, a study in the Philippines suggests that disaster initiatives tended to be driven by funding agencies or donors. There is not enough effort being taken to utilize resources within communities or empower community members by engaging them in making decisions that affect them. Such participatory processes are important to provide communities a sense of control and enable their psychological recovery (Santiago, Manuela, Tan, Sanez, & Tong, 2018). A study of Asian indigenous communities by Chua, Kadirvelu, Yasin, Choudry, and Parl (2019) reported a key factor to their resilience as their interconnectedness with their land and culture. Indigenous communities that are able to survive disasters are those that emphasize community and social cohesion and strive to keep and pass on their culture, traditions, and indigenous knowledge. The authors highlighted the importance of respecting, preserving, and harnessing this interconnectedness when intervening in communities affected by disasters.

Understanding a community's social networks is also important in planning of evacuation centers and camp management. Given the importance of social support, community disaster risk reduction and management (DRRM) plans can be crafted in such a way that families, clans, or even community groups are able to stay together and plan response and recovery actions together (Labaria et al., 2020). They also suggested the importance of including in DRRM teams social clubs and religious groups that play an influential role in community dynamics. On a positive note, there has been increasing use of participatory and development-oriented approaches in partnership with people's organization to enable responsive and sustainable disaster responses (Holden, Knadeau, & Porio, 2017).

Beyond community, organizations can also play a role in enabling collective resilience among their employees. Teng-Calleja, Sabile, and Villasanta (2020) described companies in Thailand and the Philippines that monitor and assess risks factors (e.g., rainfall and floodwater levels) and take proactive measures to protect their facilities from possible damages. These organizations invest in information and communication systems to monitor employees' situation as soon as disasters happen. They also create active employee databases that will allow them to reach out to all employees. Some companies create flood plans to reduce the time needed to recover submerged equipment and minimize disruption in their supply chain as well as ensure the well-being and safety of employees. Others prepare communication contingency equipment such as satellite phones and radio systems. They form and train emergency response teams and put in place DRRM protocols and business continuity plans. Some even allocate resources for a calamity fund, invest in rescue vehicles, identify assembly areas, and partner with other volunteer organizations and government organizations to implement safety exercises, drills, first-aid training, and support in rescue efforts. One organization has even partnered with a drugstore and a supermarket to provide medicine and food in case of disasters (Teng-Calleja et al. (2020).

In addition to these efforts, Teng-Calleja et al. (2020) described efforts of companies to provide emergency go-bags, hard hats, smoke masks, and other fire protection equipment to employees. These employees also had regular practice drills or dry-runs of systems and protocols for different types of emergencies (e.g., earthquake, fire, bomb threats). There were also those provide avenues for storytelling and ideation exercises to prepare and empower their employees to co-create disaster management initiatives.

BEYOND DISASTERS: THE ROLE OF POVERTY ALLEVIATION AND HUMAN DEVELOPMENT

For most of this chapter, we have focused on insights on disaster preparedness, resilience, and recovery. However, we would be remiss if we fail to point out a source of vulnerability to disasters in SEA. Poverty is a major contributor of vulnerability because the poor are more likely to settle in and work in areas exposed to potential hazards. They live in fragile houses and are more likely to be displaced. They are less likely to have the funds, secure employment, or access to credit or insurance that would provide them the resources to cope when a disaster strikes. They depend on natural capital as a source of food and income that are more climate-sensitive (Rubin, 2014).

Educational inequalities also have important implications for disaster risk reduction. Impoverished and marginalized communities are more likely to reside in high-risk areas and as such, have less resources to prepare for hazardous event and are more vulnerable to environmental hazards. Particularly vulnerable are those in rural areas which are more likely to have less access to education, more exposed to natural hazards, and depend more on the environment in their livelihoods (Hoffmann & Blecha, 2020). There is evidence from the Philippines and Thailand that education level is positively correlated with the probability of a household undertaking precautionary measures and being prepared for disasters (Hoffmann & Muttarak, 2017; Muttarak & Pothisiri, 2013). Hoffmann & Blecha (2020) highlighted the importance of providing universal access to formal education as a way to contribute to disaster risk reduction. They also suggested that in addition to formal and non-formal education, disaster risk reduction trainings can also help counteract existing gaps and mitigate some of the challenges and risks faced by marginalized populations. They also highlighted the importance of considering social factors such as gender and age. Beyond the vulnerability of the elderly, in SEA, women above 65 years old have a disproportionally lower literacy compared to men, making them particularly vulnerable to environmental hazards (Hoffmann & Blecha, 2020).

Urbanization also appears to be an important risk factor. A cross-province study in Vietnam reported that a high rate of urbanization, more than population density, is a strong determinant of disaster fatalities (Rubin, 2014). According to the Human Cities Coalition (2016), the steep rise in urbanization has led to increased congestion, inequality, crime, and unemployment. Moreover, the lack of information on and provision of basic services make the urban poor vulnerable to environmental health risks. All these suggest that unless governments do not address these more deeply rooted human development issues, SEA will continue to be vulnerable to climate-induced disasters.

REFERENCES

Agbayani, B. E., Villaflor, P., Villaret, N., Hechanova, M. R. (2019). The role of Filipino masculine ideology on the adaptive coping, psychological wellbeing and vicarious trauma of first responders. *International Journal of Culture and Mental Health, 11*(4), 753–762.

Alexander, D. A., & Klein, S. (2009). First responders after disasters: A review of stress reactions, at-risk, vulnerability, and resilience factors. *Prehospital and Disaster Medicine, 24*, 87–94. https://doi.org/10.1017/S1049023X00006610

Allen, K. M. (2006). Community-based disaster preparedness and climate adaptation: Local capacity-building in the Philippines. *Disasters, 30*(1), 81–101. doi:10.1111/j.1467-9523.2006.00308.x

Aten, J., O'Grady, K., Milstein, G., Boan, D., & Schruba, A. (2014). Spiritually oriented disaster psychology. *Spirituality in Clinical Practice, 1*(1), 20–28. doi:10.1037/scp0000008

Chua, R., Kadirvelu, A., Yasin, S., Choudry, F., & Parl, M. S. (2019). The cultural, family and community factors for resilience in Southeast Asia communities: A systematic review. *Journal of Community Psychology, 47*, 1750–1771.

De Perez, E., Nerlander, L., Monasso, F., Van Aalst, M., Mantilla, G., & Muli, E. (2015). Managing health risks in a changing climate: Red Cross operations in East Africa and Southeast Asia. *Journal of Climate and Development, 7*(3), 197–207.

Guilaran, J., & Nguyen, H. A. (2020). Mental health and psychosocial support services for disaster responders in Southeast Asia. In M. R. Hechanova & L. C. Waelde (Eds.), *Resistance, resilience, and recovery from disasters: Perspectives from Southeast Asia* (Community, Environment and Disaster Risk Management, Vol. 21, pp. 117-141). Bingley: Emerald Publishing.

Hechanova, M. R. & Waelde, L. (2017). The Influence of Culture on Disaster Mental Health and Psychosocial Support Interventions in Southeast Asia. *Mental Health, Religion & Culture, Published online July* 5, 2017. DOI: 10.1080/136746762017.1322048

Hechanova, M. R., Docena, P. S., Alampay, L. P., Acosta, A. C., Porio, E. E., Melgar, I. E., & Berger, R. (2018). Evaluation of a resilience intervention for Filipino displaced survivors of Super Typhoon Haiyan. *Disaster Prevention and Management*, *27*(3), 346–399. doi:10.1108/DPM-01-2018-0001

Hechanova, M. R., Waelde, L. C., Docena, P. S., Alampay, L. P., Alianan, A. S., Flores, M. J., & Melgar, I. E. (2015). The development and initial evaluation of Katatagan: A resilience intervention for Filipino disaster survivors. *Philippine Journal of Psychology*, *48*(2), 105–131.

Hechanova, M. R., Waelde, L. C., & Ramos, P. A. (2016). Evaluation of a group-based resilience intervention for typhoon Haiyan Survivors. *Journal of Pacific Rim Psychology*, *10* (e12), 1–10. doi:10.1017/prp.2016.9

Hoffmann, R., & Blecha, D. (2020). Mental health and psychosocial support services for disaster responders in Southeast Asia. In M. R. Hechanova & L. C. Waelde (Eds.), *Resistance, Resilience and Recovery from Disasters: Perspectives from Southeast Asia* (Community, Environment and Disaster Risk Management, Vol. 21, pp.17-38). Bingley: Emerald Publishing.

Hoffmann, R., & Muttarak, R. (2017). Learn from the past, prepare for the future: Impacts of education and experience on disaster preparedness in the Philippines and Thailand." *World Development*, *96*, 32–51.

Holden, W., Knadeau, K., & Porio, E. (2017). *Ecological liberation theology: Faith-based approaches to poverty and climate change in the Philippines*. New York, NY: Springer Brief in Geography.

Human Cities Coalition. (2016). Human cities coalition final report. Retrieved from https://www.humancities.co/app/uploads/2018/11/Human-Cities-Coalition-final-report-inception-phase.pdf

Inter-Agency Standing Committee. (2009). IASC guidelines on mental health and psychosocial support in emergency settings. Retrieved from https://www.refworld.org/docid/4e97ceca2.html

Ismail, N., Suwannapong, N., Howteerakul, N., Tipayamongkholgul, M., & Apinuntavech, S. (2016). Assessing disaster preparedness and mental health of community members in Aceh, Indonesia: A community-based, descriptive household survey of a national program. *Rural & Remote Health*, *16*(4), 3788.

Karanci, A. N., Aksit, B., & Dirik, G. (2005). Impact of a community disaster awareness training program in Turkey: Does it influence hazard-related cognitions and preparedness behaviors. *Social Behavior and Personality: An International Journal*, *33*, 243–258. doi:10.2224/sbp.2005.33.3.243

Usher, K, Mills, J., West, C., Casella, E., Dorji, P., Guo, A., … Woods C. (2015). Cross-sectional survey of the disaster preparedness of nurses across the Asia-Pacific region. *Nursing & Health Sciences*, *17*, 434–443.

Labaria, Acosta, & Gotangco. (2020). Mental health and psychosocial support services for disaster responders in Southeast Asia. In M. R. Hechanova & L. C. Waelde (Eds.), *Resistance, Resilience and Recovery from Disasters: Perspectives from Southeast Asia* (Community, Environment and Disaster Risk Management, Vol. 21, pp.53-74). Bingley: Emerald Publishing.

Landoy, B., Hechanova, M. R., Ramos, P. A., & Kintanar, N. (2015). The application and adaptation of Psychological First Aid: The Filipino psychologists' experience after Typhoon Haiyan. *Philippine Journal of Psychology*, *48*(2), 81–104.

Manaois, Tabo-Corpuz, & Heise. (2020). Psychological first aid: Application and Adaptation in Southeast Asia. In M. R. Hechanova & L. C. Waelde (Eds.), *Resistance, Resilience and Recovery from Disasters: Perspectives from Southeast Asia* (Community, Environment and Disaster Risk Management, Vol. 21, pp.75-94). Bingley: Emerald Publishing.

Muttarak, R., & Pothisiri, W. (2013). The Role of Education on Disaster Preparedness: Case Study of 2012 Indian Ocean Earthquakes on Thailand's Andaman Coast. *Ecology and Society*, *18*(4), 51.

Office of Civil Defense. (2011). National disaster risk reduction and management plan (NDRRMP) 2011-2028. Retrieved from https://reliefweb.int/sites/reliefweb.int/files/resources/Signed_NDRRMP.pdf

Plan International & Australian Aid. (2019). Child centred climate change adaptation (4CA), act to adapt, the next generation leads the way. Retrieved from https://plan-international.org/publications/act-adapt-child-centred-climate-change-adaptation

Quarantelli, E. L. (1999). Disaster related social behavior: Summary of 50 years of research findings. Retrieved from http://dspace.udel.edu/handle/19716/289

Rahman, A., Hamdani, S., Awan, N., Bryant, R., Dawson, K., Khan, M., … Van Ommeren, M. (2016). Effect of a multicomponent behavioral intervention in adults impaired by psychological distress in a conflict-affected area of Pakistan. Journal of the American Medical Association, *316*, 2609–2617. doi:10.1001/jama.2016.17165

Rich, G. J. & Sirikantroporn, S. (2020). Posttraumatic Growth and Resilience in Southeast Asia. In M. R. Hechanova & L. C. Waelde (Eds.), *Resistance, Resilience and Recovery from Disasters: Perspectives from Southeast Asia* (Community, Environment and Disaster Risk Management, Vol. 21, pp.143–159). Bingley: Emerald.

Rubin, O. (2014). Social vulnerability to climate induced natural disasters: Cross-provincial evidence from Vietnam. *Asia Pacific Viewpoint*, *55*, 67–80.

Samphantharak, K. (2014). Natural disasters and the economy: Some recent experiences from Southeast Asia. *Asian-Pacific Economic Literature*, *28*, 33–51. doi:10.1111/apel.12075

Santiago, J. S., Manuela, W. S., Tan, M. L., Sanez, S. K., & Tong, A. Z. (2018). Agency-driven post-disaster recovery: A comparative study of three Typhoon Washi resettlement communities in the Philippines. *International Journal of Disaster Risk Reduction*, *27*, 480–489.

Seyle, D. C., Widyatmoko, C. S., & Silver, R. C. (2013). Coping with natural disasters in Yogyakarta, Indonesia: A study of elementary school teachers. *School Psychology International*, *34*, 387–404. doi:10.1177/0143034312446889

Teng-Calleja, Sabile, & Villasanta. (2020). Enabling Organizational Resilience midst Disasters: Experiences of Work organizations in Southeast Asia. In M. R. Hechanova & L. C. Waelde (Eds.), Resistance, *Resilience and Recovery from Disasters: Perspectives from Southeast Asia* (Community, Environment and Disaster Risk Management, Vol. 21, pp. 39-52). Bingley: Emerald Publishing.

van Griensven, F., Chakkraband, M. L. S., Thienkrua, W., Pengjuntr, W., Cardozo, B. L., Tantipiwatanaskul, P., ... Tappero, J. W. (2006). Mental health problems among adults in tsunami-affected areas in Southern Thailand. Journal of the American Medical Association, 296, 537–548. doi:10.1001/jama.296.5.537

Waelde, L. C., Hechanova, R. M., Ramos, P. A. P., Macia, K. S., & Moschetto, J. M. (2018). Mindfulness and mantra training for disaster mental health workers in the Philippines. *Mindfulness*, *9*, 1181–1190. doi:10.1007/s12671-017-0855-2

World Health Organization (WHO). (2011). *Mental health atlas 2011: Indonesia.* Retrieved from https://www.who.int/mental_health/evidence/atlas/profiles/idn_mh_profile.pdf

World Health Organization (WHO). (2016). *Problem Management Plus (PM+): Individual psychological help for adults impaired by distress in communities exposed to adversity.* (Generic field-trial version 1.0). Geneva, WHO.

Yusuf, N. R. (2016). Why Indonesia must invest in mental health. Retrieved from https://www.rappler.com/world/regions/asia-pacific/indonesia/bahasa/englishedition/148788-mental-health-illness

INDEX

Printed and bound by CPI Group (UK) Ltd, Croydon, CR0 4YY

13/09/2026

14961028-0001